Lecture Notes in Physics

New Series m: Monographs

The Editorial Policy for Monographs

The series Lecture Notes in Physics reports new developments in physical research and teaching - quickly, informally, and at a high level. The type of material considered for publication in the New Series m includes monographs presenting original research or new angles in a classical field. The timeliness of a manuscript is more important than its form, which may be preliminary or tentative. Manuscripts should be reasonably self-contained. They will often present not only results of the author(s) but also related work by other people and will provide sufficient motivation, examples, and applications.

The manuscripts or a detailed description thereof should be submitted either to one of the series editors or to the managing editor. The proposal is then carefully refereed. A final decision concerning publication can often only be made on the basis of the complete manuscript, but otherwise the editors will try to make a preliminary decision as definite as they can on the basis of the available information.

Manuscripts should be no less than 100 and preferably no more than 400 pages in length. Final manuscripts should preferably be in English, or possibly in French or German. They should include a table of contents and an informative introduction accessible also to readers not particularly familiar with the topic treated. Authors are free to use the material in other publications. However, if extensive use is made elsewhere, the publisher should be informed.

Authors receive jointly 50 complimentary copies of their book. They are entitled to purchase further copies of their book at a reduced rate. As a rule no reprints of individual contributions can be supplied. No royalty is paid on Lecture Notes in Physics volumes. Commitment to publish is made by letter of interest rather than by signing a formal contract. Springer-Verlag secures the copyright for each volume.

The Production Process

The books are hardbound, and quality paper appropriate to the needs of the author(s) is used. Publication time is about ten weeks. More than twenty years of experience guarantee authors the best possible service. To reach the goal of rapid publication at a low price the technique of photographic reproduction from a camera-ready manuscript was chosen. This process shifts the main responsibility for the technical quality considerably from the publisher to the author. We therefore urge all authors to observe very carefully our guidelines for the preparation of camera-ready manuscripts, which we will supply on request. This applies especially to the quality of figures and halftones submitted for publication. Figures should be submitted as originals or glossy prints, as very often Xerox copies are not suitable for reproduction. In addition, it might be useful to look at some of the volumes already published or, especially if some atypical text is planned, to write to the Physics Editorial Department of Springer-Verlag direct. This avoids mistakes and time-consuming correspondence during the production period.

As a special service, we offer free of charge LaTeX and TeX macro packages to format the text according to Springer-Verlag's quality requirements. We strongly recommend authors to make use of this offer, as the result will be a book of considerably improved technical quality. The typescript will be reduced in size (75% of the original). Therefore, for example, any writing within figures should not be smaller than 2.5 mm.

Manuscripts not meeting the technical standard of the series will have to be returned for improvement.

For further information please contact Springer-Verlag, Physics Editorial Department II, Tiergartenstrasse 17, W-6900 Heidelberg, FRG.

Giuseppe Morandi

The Role of Topology in Classical and Quantum Physics

Springer-Verlag

Berlin Heidelberg New York
London Paris Tokyo
Hong Kong Barcelona
Budapest

Author

Giuseppe Morandi
Dipartimento di Fisica, Universitá di Bologna
CISM and INFN, Sezione di Bologna
v. Irnerio 46, I-40126 Bologna, Italy

PHYs
Sep/ae   *0 4646307*

ISBN 3-540-55088-7 Springer-Verlag Berlin Heidelberg New York
ISBN 0-387-55088-7 Springer-Verlag New York Berlin Heidelberg

Typesetting: Camera ready by author
Printing and binding: Druckhaus Beltz, Hemsbach/Bergstr.
58/3140-543210 - Printed on acid-free paper

SD   4/6/92 pl

PREFACE.

Topology entered physics first in 1931, when P.A.M. Dirac discussed, in a famous paper [35] (see also [14,15,36]), the compatibility of the existence of magnetic poles with the conceptual framework of quantum mechanics. This led Dirac to the discovery of his famous "quantization condition": $q \cdot g = n\hbar/2$, where q is the electric charge of a particle orbiting around a monopole of strength -g-. Dirac's quantization condition is the first instance in Physics of "topological quantization", i.e. of a quantization condition for a *classical* parameter forced upon us by the requirement of consistency with quantum mechanics, and arising entirely from topology.

At about the same time, H. Hopf discovered the fibration $S^1 \to S^3 \to S^2$ [26]. That the two structures were actually intimately related became clear only more than thirty years later, when it was realized that the fiber bundle corresponding to the Hopf fibration can be endowed with a *natural connection* [25] whose curvature can be identified with the field of a magnetic monopole sitting at the center of the sphere S^2.

Another instance in which nontrivial topological properties of space-time appear to play a relevant role is provided by the effect discussed by Y. Aharonov and D. Bohm in 1959 [3,75,76,79,82,88]. Although discussed originally as a scattering event, the effect can also be described in different terms   by saying that the wave function of a charged particle which is adiabatically dragged around an infinitely long solenoid enclosing a flux Φ acquires an extra phase of $\exp[2\pi i\Phi/\Phi_0]$, where $\Phi_0 = hc/q$. For physicists, this whole set of ideas was first exposed in a systematic way in 1975 in an influential paper by T.T. Wu and C.N. Yang [110]. What these authors stressed was that the proper language to describe quantum mechanics in the presence of electromagnetic couplings is that of $U(1)$ principal fiber bundles [30,60,94], and that wave functions are to be properly seen as *sections* [30] of such bundles. This paved the way to the development of gauge field theories.

In the case of the Aharonov Bohm effect, the bundle is flat, but has a nontrivial holonomy, and the phase acquired by the wave function is just a manifestation of the holonomy [30] of the bundle.

In the recent past, starting from a seminal paper published by M.V. Berry in 1984 [18], a flurry of activity has taken place around a complex of physical ideas that go under the name of "quantum adiabatic phase", or the "Berry phase", all sharing the common mathematical theme of holonomy [4,5,19,21,22,27,86,88,89,108,114]. Quite remarkably, Berry's phase was found to have a precursor also in a classical context. Actually, J. Hannay showed [53] that similar holonomy (or, better, *an*-holonomy) effects can also show up in purely classical (integrable) dynamical systems subject to an adiabatic excursion in some parameter space. Another, and even more remarkable, precursor of Berry's phase had actually been found, back in 1956, by S. Pancharatnam [20,81] in the realm of optics of plane polarized light. This topological effect takes place in a purely classical context as well.

The Aharonov-Bohm effect is also a prototype of the kind of problems that arise when one tries to enforce Quantum Mechanics in nonsimply connected spaces [12-15,37,56,58,59,71,93, 96].

The study of the latter began in a systematic way in the early 1970s, with a pioneering paper by M.G.C. Laidlaw and C. Morette-deWitt [67]. Since this first investigation, it has become clear that, in nonsimply connected spaces, one deals with *genuinely inequivalent Quantum Mechanics,* and that the latter are "indexed", in the scalar case, by the first homology group with integer coefficients [54,57] of the configuration space Q, $H_1(Q,\mathbb{Z})$ and, more generally, by the first homotopy group $\pi_1(Q)$.

A conspicuos example of quantization on nonsimply connected spaces is provided by the case of identical particles moving in $\mathbb{R}^2$, in which case: $\pi_1(Q)=\mathbb{B}_N$ for N particles, where $\mathbb{B}_N$ is the so-called N-string braid group [9,23,50,56], whose one-dimensional unitary representations are indexed by an angle θ, with $\theta=0$ corresponding to bosons, $\theta=\pi$ to fermions. For generic θ, F. Wilczek coined the name "anyons" [104-107] for identical particles obeying statistics which are neither Bose nor Fermi [6,8,46,56,59,95,111-113]. "Anyon statistics" can be dynamically implemented if the particles (which can be "nominal" bosons or fermions to start with) interact with a gauge field whose field Lagrangian is dominated by a topological term, the so-called *Chern-Simons* term [29,48,61-64,116].

That two-dimensional electron systems are not simply an amusing academic game, but can be relevant for physics, was shown first in the study of the Fractional Quantum Hall Effect (FQHE) [8,51,69, 84,88,108,113]. Indeed, there seems to be growing (theoretical, at least) evidence that elementary excitations in the FQHE obey fractional statistics.

Finally, anyonic excitations have been advocated [16,17,27,70,106,107] also as a possible mechanism to explain the unusual behavior of the newly discovered high-T_c superconducting materials.

Going back again for a moment to a purely classical context, we should also recall the scientific activity on the topological classification of defects in ordered media which started around the mid-1970s, and is associated with the names of N.D. Mermin and G. Toulouse [72,98]. In this approach, line and point defects are classified by the first and second homotopy group of the order-parameter space respectively, and these groups provide almost all of the relevant information concerning the stability and the laws of combination of such defects.

In a slightly different, but closely related, context, the topological analysis has been widely used to classify *topological* solutions of *classical* field theories [37,48,84,96]. This classification provides a useful and essential starting point for the quantization of such theories.

In these Lectures, which were delivered at the Physics Department of the University of Shanxi (Taiyuan, Shanxi Province, PRC) in Summer 1990, I discussed some of the topics listed above, with no pretention whatsoever to originality and/or completeness. Concerning the topological theory of defects, I have not really discussed it, but borrowed some of its language from the classic paper by N.D. Mermin [72], to the extent that it could be useful to introduce general concepts in algebraic topology. The clarity and beauty of Mermin's original paper are still largely unequalled, and the interested reader is referred to it for a thorough and highly readable account of the subject. Also I have not touched in these Lecture Notes upon the subject of high-T_c and anyon superconductivity, since there are by now extensive reviews [17,106] available on the subject.

Since what I am presenting here is, in the spirit of the Springer series in which it will appear, only a slightly polished and revised form of an original set of lecture notes, no attempt has been made to make them completely self-consistent nor to give a fully comprehensive list of references. As to the latter, I quoted only those references and sources I was aware of and familiar with. Authors whom I have possibly ignored should blame therefore only my personal ignorance. As to self-consistency, I did my best to achieve it but could not help assuming some prerequisites, namely that the audience (and hence, now, the readership) had some basic knowledge of elements of differential geometry, fiber bundles and connections, differential forms and deRham's cohomology. All these topics are well covered by existing and, by now, standard literature and textbooks [1,26,28,30,31,40,43, 45, 60,77,78,85,90,94,99,102]. Trying to cover them as well in this book would have risked turning it into a many-volume treatise, which was of course outside my scopes and, in particular, beyond my capabilities. However, the topics just mentioned are essentially the only prerequisites for this book. I do hope that I have succeeded at least partly in making topology and its applications understandable and available to an as large as possible audience of theoretical and mathematically-oriented physicists.

Bologna, September 1991   Giuseppe Morandi

CONTENTS.

CH. 1. AN ELEMENTARY INTRODUCTION TO ALGEBRAIC TOPOLOGY.

1.1. INTRODUCTION.

As already mentioned in the Introduction, we will use here the ideas of **ordered media** and of **order parameter(s)** for such media mainly as an excuse for introducing basic concepts of algebraic topology and especially the relevant notions concerning homotopy groups that we will exploit extensively in what follows.

An **ordered medium** will be defined as follows: Let Q be a "physical" space (e.g.: $Q=\mathbb{R}^n$ for some n) and let T be another space, the "target" space of a family of maps or, by (our) definition the "order-parameter space" (e.g.: $T= S^2$, S^3 or, more generally, any (left, but it could equally be right) coset space G/H, with G a Lie group and H a subgroup of G). Both Q and T are supposed to be endowed with (some kind of) a topology, and will be then considered as **topological spaces**. The medium is (again by definition) "ordered" when an **order parameter distribution** is given, namely a (continuous at least, but we will leisurely assume C^k (and, in fact, C^∞) for any k we may need in the following) map:

$$f: Q \to T \qquad (1.1)$$

The medium will be called **uniform** iff f=const., **nonuniform** otherwise.

Definition: A Defect is a point, a line or a surface on which the map -f- becomes singular.

Remarks:

i) It will be assumed throughout that Q and T are "decent" enough that they can be attributed a well-defined (and constant) dimension. "dim$Q$" and "dim$T$" are then meaningful and nonambiguous concepts.

ii) Defects are, broadly speaking, regions in Q of dimensionality **lower** than dim$Q$ on which the order-parameter field becomes singular.

1.2. SOME EXAMPLES.

i) Planar Spins. In this case: $Q=\mathbb{R}^2$, and $T=S^1$. The order-parameter map is defined by:

$$f: \mathbb{R}^2 \to S^1 \quad \text{by: } f(\vec{r}) = \hat{u}\,\cos\phi(\vec{r}) + \hat{v}\,\sin\phi(\vec{r}) \tag{1.2}$$

with $\vec{r} \in \mathbb{R}^2$ and $\hat{u}$, $\hat{v}$ a pair of orthonormal vectors in $\mathbb{R}^2$.

ii) Superfluid He^4 and Superconductors. Here: $Q = \mathbb{R}^3$, and the order parameter is a complex scalar field (the wavefunction of the condensate):

$$\psi(\vec{r}) = \psi_0 \exp[i\phi(\vec{r})] \tag{1.3}$$

In most cases $\psi_0 =: |\psi| = $ const., and therefore $\mathsf{T} = \mathsf{U}(1) \sim \mathsf{S}^1$.

iii) Ordinary spins. In this case, $Q = \mathbb{R}^d$ for some d (usually d=2 or d=3), and the order parameter is a **unit vector in three-space**:

$$f(\vec{r}) = \vec{s}(\vec{r}) \; ; \; \vec{s} \cdot \vec{s} \equiv 1 \qquad (1.4)$$

Hence: $\mathsf{T} = \mathsf{S}^2$, the two-sphere in $\mathbb{R}^3$.

iv) Nematic liquid crystals. In this case the order parameter is **a field of directions** (in $\mathbb{R}^3$), and hence it can be identified with a unit vector "without the arrowhead". Therefore: $\mathsf{T} = \{\mathsf{S}^2$ with identification of antipodal points\}, i.e.: $\mathsf{T} \simeq \mathbb{RP}^2$, the **real projective plane**. Another identification/description of T is as **the set of all real symmetric, unimodular and traceless** 3×3 **matrices with a doubly degenerate eigenvalue.** Indeed, it is a simple exercise to show that any such matrix $\mathcal{M}$ can be written as

$$\mathcal{M} = \lambda(s_i s_j - \tfrac{1}{3}\delta_{ij}) \qquad (1.5)$$

where: $\vec{s} \cdot \vec{s} = 1$ and λ is a normalization factor. As $\vec{s}$ and $-\vec{s}$ determine the same matrix, $\mathcal{M}$ will be determined bijectively by the **direction** of $\vec{s}$.

1.3. TOPOLOGY OF DEFECTS (THE CASE OF PLANAR SPINS).

Let's consider a system of planar spins (Ex. i) above), and suppose we take a "big" circle (i.e., topologically, a closed loop) $\mathcal{C}$ which crosses no singularities of the order parameter. As the latter is single-valued by definition, the total variation along the circle of the angle $\vec{s}(\vec{r})$ makes w.r.t. a fixed direction can only be $2\pi n$ for some integer n.

Definition: n is the *winding number* associated with the loop $\mathcal{C}$.

Fig. 1 shows some examples of spin configurations and the associated winding numbers.

If we now deform $\mathcal{C}$ continuously, n, which is a discrete quantity, cannot vary. It follows then by a simple reasoning that *the region enclosed by a loop with nonzero winding number must contain one or more singularities of the order parameter.* If there are no singularities, then n=0, but note that **the contrary need not be true.** However, one can get rid of the n=0 singularities by **purely local** modifications ("local surgery"[72]) of the order-parameter distribution. More generally, one can, again by "local surgery", modify two distributions having the same winding number in such a way that they acquire the same core without altering the field distribution at large distances. This is illustrated in Figs. 2 and 3.

All in all, this amounts to say that n=0 configurations can be deformed into completely smooth configurations by purely **local**

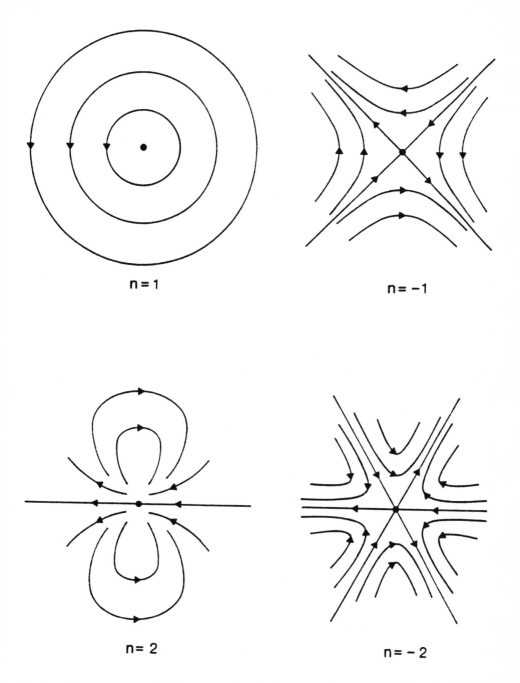

Fig.1. Examples of different point defects in the spin configurations of a system of planar spins, and the associated winding numbers.

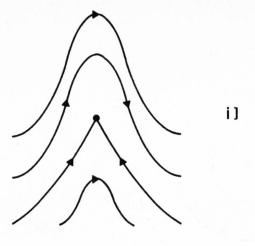

i]

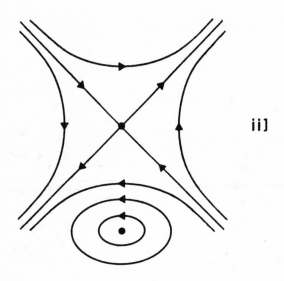

ii]

Fig.2. Examples of defects (planar spins) with zero total winding numbers: i) A single defect with n=0; ii) Two defects with opposite winding numbers.

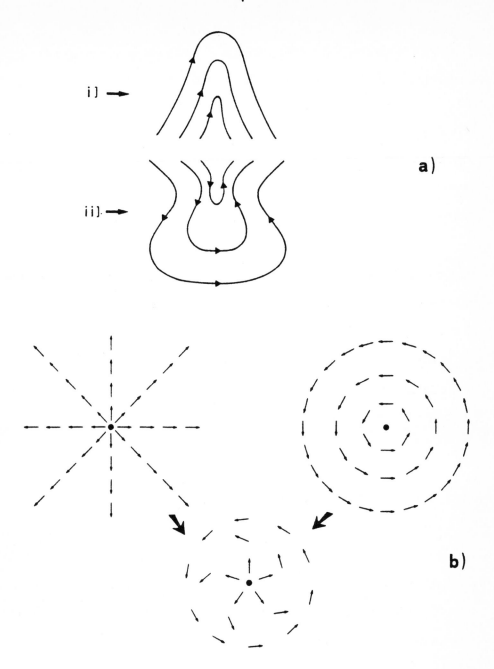

i] →

ii]. →

a)

b)

Fig.3. a) "Local surgery" operations modifying the field configurations of Fig.2 into defect-free ones. b) "Local surgery" modifying the core of two defects with the same winding number into one another.

operations, while this is not possible for n $\neq$ 0 configurations. So:

— *Defects with n $\neq$ 0 are <u>topologically stable</u> , while*
— *Defects with n=0 are <u>topologically unstable</u>*

Moreover, in general:

Defects with different winding numbers can combine into a single defect whose winding number is the sum of the winding numbers of the original defects (see Fig. 4). □. It turns out [72] that this combination law holds only iff the defects are classified by a single winding number or, more generally, iff the fundamental group of the order-parameter space (see below) is abelian.

1.4. THE FUNDAMENTAL GROUP.

Winding numbers are connected with maps of loops in **Q** into the parameter space **T**. As the former are parametrized by the circle S^1, what we are ultimately studying are *maps of S^1 into* **T** or, equivalently, maps of the form:

$$f: [0,1] \rightarrow T ; \; f(0)= f(1) \qquad (1.6)$$

Any two such maps will be called (mutually) *homotopic* iff they can be deformed continuously into each other. More formally [57,78]:

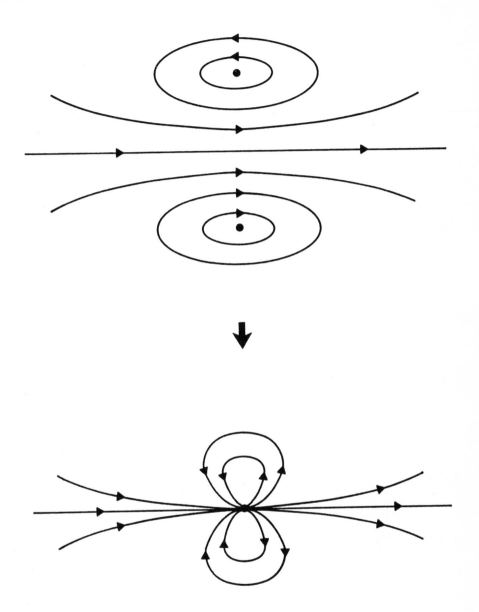

Fig.4. Illustration of the procedure of combination of abelian defects.

Definition: Two maps f and g of the form (1.6) are *homotopic* iff there is a continuous map:

$$h: [0,1] \times [0,1] \to \mathbb{T} \tag{1.7}$$

s.t. (we will write $h_t(.)$ for $h(t,.)$, $t \in [0,1]$):

$$h_0 \equiv f \; ; \; h_1 \equiv g \; ; h_t(0) = h_t(1) \; \forall \; t \tag{1.8}$$

Remarks:

i) What we are defining here is the so-called *free homotopy* of loops (the image of f in $\mathbb{T}$ being a closed path, i.e. a loop in $\mathbb{T}$). In other words, ·the loops associated with f and g need not have any common points.

ii) For every t, h_t defines a loop in $\mathbb{T}$ which continuously interpolates (as t varies) between f and g. Viceversa, for every z (in the second $[0,1]$ of (1.7)) $h_z =: h(.,z)$ defines a path in $\mathbb{T}$ from $h_z(0) = f(z)$ to $h_z(1) = g(z)$.

We want to show now that there is a group structure associated with the set of loops in $\mathbb{T}$. In order to do this, we must however study first the so-called:

BASED HOMOTOPY.

The order-parameter space $\mathbb{T}$ shall be always assumed to be **connected and path-connected** [77,78,89]. Rememeber that path-connectedness does **not** follow from connectedness. The classical example showing this [89] is the subset of $\mathbb{R}^2$ defined (with the induced topology) by:

$$\mathbb{T}= \{[-1 \leq y \leq 1] \cup [\text{Graph of: } f(x)=\sin(\tfrac{1}{x})]; \ 0 < x \leq \tfrac{\pi}{2}\} \quad (1.9)$$

which is connected but not path-connected.

Take now $x \in \mathbb{T}$. A **loop at x** is a loop f satisfying:

$$f(0)=f(1)=x \quad\quad\quad\quad (1.10)$$

and any two loops f and g at x are homotopic iff they satisfy (1.7), while the second of (1.8) is modified into the stronger condition:

$$h_t(0)= h_t(1)=x \ \ \forall \, t \in [0,1] \quad\quad (1.11)$$

Note that a (based or unbased) loop is defined by a closed curve in $\mathbb{T}$ **and** a parametrization. If we change the parametrization of a loop in a continuous way, we will obtain a new loop which can be shown in an obvious way to be homotopic to the previous one. One can also prove almost at once the following:

Theorem: Homotopy (both free and based) is an **equivalence relation**

among loops. The set of (based or unbased) loops in T can be partitioned into pairwise disjoint equivalence classes of mutually homotopic loops. $\square$.

PRODUCT OF LOOPS.

The **product** of two loops f and g is defined as:

$$(f \cdot g)(z) = \begin{cases} f(2z) , & 0 \leq z \leq \frac{1}{2} \\ \\ g(2z-1) , & \frac{1}{2} \leq z \leq 1 \end{cases} \tag{1.12}$$

Also, the **inverse** of a loop f is defined by:

$$f^{-1}(z) = f(1-z) \tag{1.13}$$

The **trivial loop** at a given point x is denoted by e, and is defined by:

$$e(z) \equiv x \quad \forall z \tag{1.14}$$

All this apparently gives the set of loops a group structure. This is however not so, as

i) Though e is a good candidate for being the identity, $f \cdot f^{-1} \neq f^{-1} \cdot f$ (because of different parametrizations) and (again for the same reason) they both differ from e.

ii) The loop product, as defined in (1.12), *is not associative*. Again because of parametrization:

$$(f \cdot g) \cdot h \neq f \cdot (g \cdot h) \tag{1.15}$$

(the proof is left as an exercise).

A group structure can nonetheless be imposed on the <u>homotopy classes</u> of loops based at any point $x \in$ T, and the argument goes as follows:

PRODUCT OF HOMOTOPY CLASSES:

Definition: Let f be a loop at $x \in$ T. The class of loops at x which are homotopic to f will be denoted by [f].

Homotopy of two loops f and g will be denoted by writing: $f \sim g$ (and we have already stated that " $\sim$ " is an equivalence relation).

Remark: [f] does *not* depend upon the parametrization of the loop as long as the latter is changed in a smooth way. Also: [f]=[g] iff $f \sim g$.

The ***product*** of two homotopy classes [f] and [g] will be defined as:

$$[f] \cdot [g] =: [f \cdot g] \tag{1.16}$$

In order that the above definition be consistent we must make sure that:

i) [f] does not depend on the choice of f within a given homotopy class. This is ensured by the very definition of [f] (see also the previous Remark).

ii) [f·g] does not depend on the choices of f and g within the respective homotopy classes, but only on the classes themselves. This is granted by the following

Theorem: If $f \sim f'$ and $g \sim g'$, then $f \cdot g \sim f' \cdot g'$.

The proof of the theorem is simple, and is left as an exercise.$\square$.

The product of homotopy classes has the following properties:

i) $[\ ] \cdot [\ ]$ is *associative*. Indeed: $[f] \cdot ([g] \cdot [h]) =: [f] \cdot [g \cdot h] = [f \cdot (g \cdot h)]$. On the other hand: $([f] \cdot [g]) \cdot [h] = [f \cdot g] \cdot [h] = [(f \cdot g) \cdot h]$. The assertion follows then from: $(f \cdot g) \cdot h \sim f \cdot (g \cdot h)$ (the two differing only by a smooth change in the parametrization. Proof left by exercise) $\square$.

ii) Existence of the *identity* and of the *inverse*.

Let [e] be the homotopy class of loops at x that are homotopic to the trivial loop. Then:

$$[e] \cdot [f] = [f] \cdot [e] = [f] \quad \forall\ [f] \tag{1.17}$$

Moreover, with the inverse loop defined as in (1.13):

$$[f] \cdot [f^{-1}] = [f^{-1}] \cdot [f] = [e] \tag{1.18}$$

Indeed, as [f] can be represented by any loop in the corresponding homotopy class, we can choose, in order to define the product $[f] \cdot [f^{-1}]$, a loop f and its inverse. The homotopy shrinking $f \cdot f^{-1}$ (as well as $f^{-1} \cdot f$) to the trivial loop is given by:

$$h_t(z) = \begin{cases} f(2zt) & 0 \leq z \leq \frac{1}{2} \\ \\ f(2t(1-z)) & \frac{1}{2} \leq z \leq 1 \end{cases} \tag{1.19}$$

(Note that: $f(2(1-z)) \equiv f^{-1}(2z-1)$, so: $h_1 \equiv f \cdot f^{-1}$). A similar homotopy can be constructed to prove that $f^{-1} \cdot f \sim e$ as well.

In conclusion, we have proved the following

Theorem:

The set of homotopy classes of loops based at any $x \in T$ can be endowed with a group structure, with the group operation being defined by:

$$[f] \cdot [g]^{-1} = [f \cdot g^{-1}] \; ; \quad [g]^{-1} = [g^{-1}] \tag{1.20}$$

□

Definition: *The group just defined is called the fundamental group at x.*

and is denoted as $\pi_1(T,x)$.□

EXAMPLES.

i) $T=S^1$. It is a simple exercise to show that:

$$\pi_1(S^1,x)= \mathbb{Z} \ , \ \forall x \in S^1 \tag{1.21}$$

ii) $T=S^2$. Then:

$$\pi_1(S^2,x)= 0 \ , \ \forall x \in S^2 \tag{1.22}$$

To show this, one has to show that any loop in S^2 can be shrunk to a point. The easiest way to prove this statement is, given a loop f, to "punch a hole" in the sphere at any point not on f and then to stereographically project the sphere onto $\mathbb{R}^2$. But $\mathbb{R}^2$ is a vector space, and any such space is always contractible to a point. Stated otherwise, if: $\vec{x}=\vec{x}(z)$ is a loop (in $\mathbb{R}^n$ for any n), it can be shrunk to a point by the homotopy: $\vec{h}_t(z)=: t\vec{x}(z)$.

iii) The "Figure-eight". This is an almost paradigmatic example [84] of a space exhibiting (at any point) a **nonabelian** fundamental group. With reference to Fig. 5, we see that

$$f \text{ is not homotopic to g, but } c \cdot f \cdot c^{-1} \sim g \tag{1.23}$$

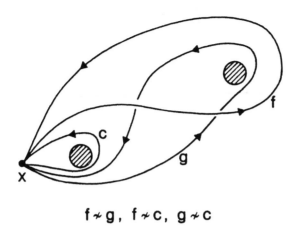

$f \nsim g, \quad f \nsim c, \quad g \nsim c$

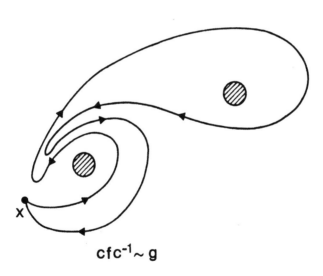

$cfc^{-1} \sim g$

Fig.5. Non homotopic (based) loops for the "figure-eight", illustrating the fact that the latter has a non abelian fundamental group.

Now, the last homotopy relation in (1.23) implies $c \cdot f \sim g \cdot c$. If the group were abelian: $c \cdot f \sim f \cdot c \sim g \cdot c$ would imply $f \sim g$, which we know to be false. $\square$.

It can be shown that the fundamental group of the figure-eight is an infinite nonabelian group, and precisely the free group [84] on two generators. We recall that a free group on n generators can be defined formally as follows: take n symbols (the "generators") $a_1, ..., a_n$ plus their inverses and the identity. Then, the group is the set of all strings of symbols (also called "words") of arbitrary length, represented as: $a_i \cdot a_j \cdot ... a_k \cdot ... a_l$. Given any two words α and β, the product $\alpha \cdot \beta$ is the word obtained by writing the two strings of symbols in sequence. It is easy to show (and it is left as an exercise) that in this way the n generators do indeed generate a group, and that the latter is always an infinite group. The choice of the generators is highly non-unique. Each choice gives rise to what is called a (different) "presentation" [84] of the group itself. In the case of the figure-eight, the two generators can be identified easily with the two independent loops in the figure.

1.5. THE ABSTRACT FUNDAMENTAL GROUP. CONJUGACY
CLASSES AND FREELY HOMOTOPIC LOOPS.

As $\mathbb{T}$ is path-connected, given any two points x and y there is a path:

$$c: [0,1] \rightarrow \mathbb{T} , \quad c(0)=x, \quad c(1)=y \tag{1.24}$$

joining them, while $c^{-1}(z) =: c(1-z)$ does the same job in the reverse order. If then f is a loop based at y:

$$c \cdot f \cdot c^{-1}: [0,1] \to \mathbb{T} , c \cdot f \cdot c^{-1}(0) = c \cdot f \cdot c^{-1}(1) = x \qquad (1.25)$$

is a loop based at x, and viceversa. So, we have succeeded in establishing *a bijection between the sets of loops based at different points in* **T**. This can be extended to an *isomorphism between the corresponding (based) homotopy groups* as follows:

i) Let $f \sim g$ at y, and let h_t be the corresponding homotopy. Then: $c \cdot f \cdot c^{-1} \sim c \cdot g \cdot c^{-1}$ at x, the homotopy (at x) being provided by $c \cdot h_t \cdot c^{-1}$. It follows that the path c defines a map (actually a bijection) of homotopy classes at y into homotopy classes at x via:

$$c \, [f] =: [c \cdot f \cdot c^{-1}] \qquad (1.26)$$

It is also almost trivial to prove that (again the proof is left as an exercise):

$$c[f] = c[g] \Leftrightarrow c \cdot f \cdot c^{-1} \sim c \cdot g \cdot c^{-1} \Leftrightarrow f \sim g \qquad (1.27)$$

ii) The map c is *compatible with the group structure* of the based homotopy groups. Indeed:

$$c([f] \cdot [g]) = c[f \cdot g] = [c \cdot (f \cdot g) \cdot c^{-1}] = [c \cdot f \cdot c^{-1} \cdot c \cdot g \cdot c^{-1}] =$$

$$= [c \cdot f \cdot c^{-1}] \cdot [c \cdot g \cdot c^{-1}] = (c[f]) \cdot (c[g]) \quad \square \qquad (1.28)$$

It follows then that, $\forall$ x,y $\in \mathbb{T}$, $\pi_1(\mathbb{T},x)$ is *isomorphic* to $\pi_1(\mathbb{T},y)$. Hence, there is *a single abstract group*, which we will call from now on $\pi_1(\mathbb{T})$, of which the based fundamental groups are isomorphic copies.

Definition: $\pi_1(\mathbb{T})$ is called the *first homotopy group*, or the *fundamental group* of $\mathbb{T}$. The topological space $\mathbb{T}$ will be called *simply connected* iff $\pi_1 = 0$ (by "0" we will denote from now onwards the trivial group consisting of the identity alone).

In order to establish the above-discussed path isomorphism we had to choose a path. The question then arises whether or not the isomorphism thus established is "canonical" or not, i.e. whether or not it depends on our choice of the path. The answer is contained in the following

Theorem: The path isomorphism between based fundamental groups is canonical iff $\pi_1(\mathbb{T})$ is abelian.

Sketch of the proof: If the isomorphism is canonical, then (Fig. 6-a):

$$c_1 \cdot f \cdot c_1^{-1} \sim c_2 \cdot f \cdot c_2^{-1} \Rightarrow h \cdot f \cdot h^{-1} \sim f \, , \ h =: c_2^{-1} \cdot c_1^{-1} \qquad (1.29)$$

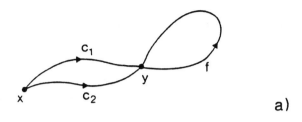

a)

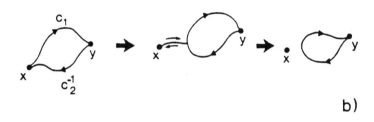

b)

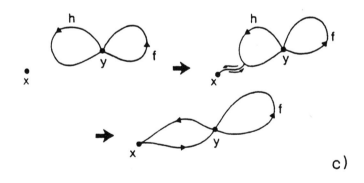

c)

Fig.6. a) Two paths c_1 and c_2 used to establish the path-isomorphism between $\pi_1(T,x)$ and $\pi_1(T,y)$. b) "Disentangling" the loop h from x in a homotopic way. c) The opposite operation, converting homotopically a loop at y into a loop passing also through x.

But h is a loop at y, and so (1.29) implies

$$h \cdot f \sim f \cdot h \tag{1.30}$$

The fact that h is a loop at y but also through x is of no importance. Indeed, h can be "disentangled" from x in a homotopic manner in the way indicated in Fig. 6-b). Conversely, any loop h at y can be converted (in a homotopic manner) into a loop (through x) of the form $c_1^{-1} \cdot c_2$, as indicated in Fig. 6-c). It follows that (1.30) holds for **any** loop h at y, and hence $\pi_1(\mathbb{T})$ is abelian. Conversely, if $\pi_1(\mathbb{T})$ is abelian, and hence (1.30) holds for any two loops, it of course implies $h \cdot f \cdot h^{-1} \sim f$ for $h = c_2^{-1} \cdot c_1$, and hence the first of (1.29).$\square$

Let now $\pi_1(\mathbb{T})$ be non-abelian. Then there are at least two paths, c_1 and c_2, such that $c_1 \cdot f \cdot c_1^{-1}$ is not homotopic to $c_2 \cdot f \cdot c_2^{-1}$ for at leats one loop f at y. But:

$$c_1 \cdot f \cdot c_1^{-1} \sim k \cdot (c_2 \cdot f \cdot c_2^{-1}) \cdot k^{-1} , \quad k =: c_1 \cdot c_2^{-1} \tag{1.31}$$

and k is a loop based at x. In terms of homotopy classes:

$$[c_1 \cdot f \cdot c_1^{-1}] = [k] \cdot [c_2 \cdot f \cdot c_2^{-1}] \cdot [k]^{-1} \tag{1.32}$$

Therefore, the two elements of $\pi_1(\mathbb{T}.x)$ onto which $[f] \in \pi_1(\mathbb{T},y)$ is mapped belong to the same conjugacy class in $\pi_1(\mathbb{T},x)$. Hence, we have the following:

Theorem: The path isomorphism establishes a canonical bijection between the conjugacy classes of the based fundamental groups. $\square$.

We can apply this result to the study of freely homotopic loops. For the latter, we have the following

Theorem: Two loops f and g (which need not have any common points) are freely homotopic iff there is a path

$$c: [0,1] \rightarrow \mathbb{T} \ , \ c(0)=x \in f, \ c(1)=y \in g \tag{1.33}$$

such that f is homotopic at x to $c \cdot g \cdot c^{-1}$.

Indeed (see Fig. 7) if the path c exists, then g is freely homotopic to $c \cdot g \cdot c^{-1}$, the homotopy being given by a continuous retraction of c and c^{-1} to a point. The loop f being in turn homotopic to $c \cdot g \cdot c^{-1}$, f and g are freely homotopic. Viceversa, let f and g be freely homotopic, and let h_t be a homotopy connecting them. Let c be the path traced by $h_t(0)$ as t ranges from 0 to 1. The homotopy between f and $c \cdot g \cdot c^{-1}$ is given by the following construction:
Define

$$c_t(z)=: h_{tz}(0) \ ; \ c_1(z) \equiv c(z) \tag{1.34}$$

Then:

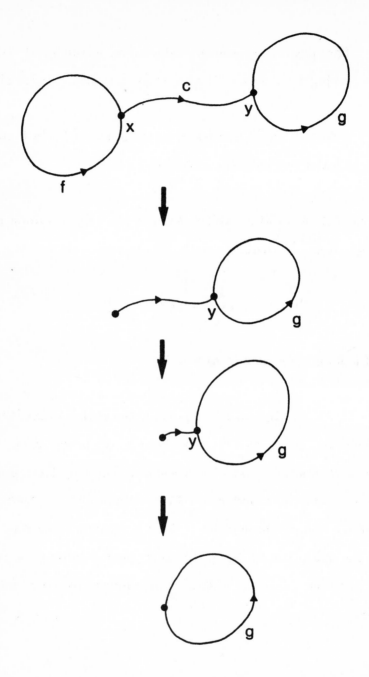

Fig. 7. Illustration of the procedure used to prove the first part of the theorem of p. 23.

$$c_t(0)= h_0(0)= x \;\; ; \;\; c_t(1)= h_t(0) \;\; ; \;\; c_0(z) \equiv h_0(0)= x \qquad (1.35)$$

Next, define:

$$k_t(z)= \begin{cases} c_t(4z/t) \,, & 0 \le z < t/4 \\ h_t(\frac{4z-t}{4-3t}) \,, & t/4 \le z \le 1-t/2 \\ c_t(2\frac{1-z}{t}) \,, & 1-t/2 < z \le 1 \end{cases} \qquad (1.36)$$

It follows from (1.36) that:

$$k_t(0)= k_t(1)= c_t(0)= x \qquad (1.37)$$

Moreover, since

$$h_t(\frac{4z-t}{4-3t})\,|_{z=t/4}= h_t(0)= c_t(1)= c_t(4z/t)\,|_{z=t/4} \qquad (1.38)$$

and

$$h_t(\frac{4z-t}{4-3t})\,|_{z=1-t/2}= h_t(1)= h_t(0)= c_t(1)= c_t(2\frac{1-z}{t})\,|_{z=1-t/2} \qquad (1.39)$$

it follows that $k_t(z)$ is a continuous map. Finally:

$$k_0(z) \equiv h_0(z)= f(z) \;\; ; \;\; k_1(z)= (c \cdot g) \cdot c^{-1} \qquad (1.40)$$

But $(c \cdot g) \cdot c^{-1}$ is a representative of the homotopy class $[cgc^{-1}]$, so k_t is the required homotopy between f and $c \cdot g \cdot c^{-1}$. This achieves the proof of the theorem.□

The previous theorem implies the following:

Theorem: Two loops at $x \in \mathbb{T}$ are freely homotopic iff they are in the same conjugacy class.

Indeed, if the two loops are freely homotopic, the previous construction applies. But now $x=y$, and c becomes a loop. Hence, [f] and [g] are in the same conjugacy class. Conversely, if [f] and [g] are in the same conjugacy class, i.e. $f \sim b \cdot g \cdot b^{-1}$ for a loop b, then the construction of the free homotopy connecting them is depicted in Fig. 8.$\square$.

From the very definition of the path isomorphism, we can also restate the second theorem of p. 22 by saying that *a loop f at x and a loop g at y are freely homotopic iff there is a path isomorphism taking the homotopy class [g] of $\pi_1(\mathbb{T},y)$ into the homotopy class [f] of $\pi_1(\mathbb{T},x)$.* $\square$.

Remembering that any two (based) loops in the same conjugacy class are freely homotopic and that free homotopy is an equivalence relation, we immediately conclude that if a loop f at x is freely homotopic to a loop g at y, *so are any pairs of loops in the corresponding conjugacy classes.* Therefore, *classes of freely homotopic loops are labeled by the conjugacy classes of $\pi_1(\mathbb{T})$.* It is only when $\pi_1(\mathbb{T})$ is abelian that classes of freely homotopic loops are directly labeled by individual elements of $\pi_1(\mathbb{T})$.

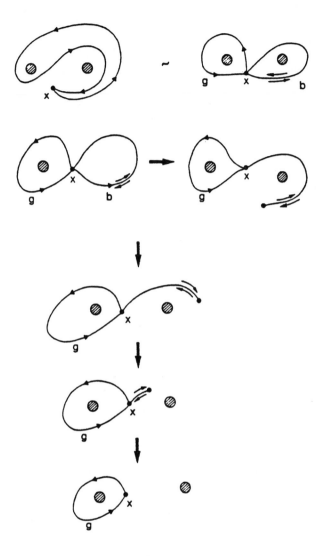

Fig. 8. Construction of the free homotopy for the theorem of p. 25. The upper part of the figure shows the homotopy between f and $b \cdot g \cdot b^{-1}$. In the lower part, as the loop which shrinks to g is not based at x, the associated homotopy is a free homotopy.

1.6. THE ORDER PARAMETER SPACE AS A COSET SPACE.

Let us begin by recollecting some elementary notions of group theory. What we will be mainly concerned with will be Lie groups, although most of what we will say applies to topological (not necessarily Lie) groups as well.

Let G be a group, and let $G_0 \subset G$ be a (proper) subgroup. Then:

i) A *left coset* of G_0 associated with the element $a \in G$ is the set:

$$a G_0 = \{ah \; ; \; h \in G_0\} \qquad (1.41)$$

Right cosets are defined in a similar way. Two (left) cosets are either disjoint or coincident. Indeed, let $g \in aG_0 \cap bG_0$. Then:

$$g = ah = bh' \quad \text{for some } h, h' \in G_0 \qquad (1.42)$$

But then $a^{-1}b \in G_0$ and

$$bG_0 = a(a^{-1}b)G_0 \equiv aG_0 \qquad (1.43)$$

$\square$

(equivalently, one could say that "to belong to a given coset" is an equivalence relation in G).

Going to the quotient, we can construct the set G/G_0 of (left or right) cosets.

ii) G_0 will be called a **normal** subgroup iff

$$aG_0 = G_0 a \quad \forall \, a \in G \tag{1.44}$$

i.e. iff left and right cosets coincide for every $a \in G$.

iii) If G_0 is normal, then **G/G_0 is itself a group**, with G_0 acting as the identity, the product being defined by:

$$(aG_0) \cdot (bG_0) = a(G_0 b)G_0 = (ab)G_0 \quad \square \tag{1.45}$$

All the above applies to any group. Let now G be a Lie (resp., topological) group, and let G_0 be **the connected component of the identity of G.** Then:

Theorem: G_0 is a normal subgroup, and G/G_0 has as its elements the connected components of G.

Being a normal subgroup is also expressed by

$$aG_0 a^{-1} = G_0 \tag{1.46}$$

For G_0 to be a group, we must first prove that ab^{-1} is in G_0 whenever a and b are. But, G_0 being connected with the identity $e \in G$, there are continuous paths a_t and b_t such that $a_0 = b_0 = e$, $a_1 = a$, $b_1 = b$. Hence, the path $a_t b_t^{-1}$ connects e with ab^{-1}, and hence $ab^{-1} \in G_0$. That $aha^{-1} \in G_0$ follows again from the existence of the path h_t, $h_0 = e$, $h_1 = b$. Then, $ah_t a^{-1}$ connects e with aha^{-1}. Hence $aha^{-1} \in G_0$ and G_0 is normal.$\square$

Any left (right) coset of G_0 is connected. Indeed, let ah, ah' $\in$ aG_0. As h,h' $\in$ G_0, there is a path b_t, b_0=h, b_1=h' connecting them in G_0. But then $ab_t \in$ aG_0 $\forall$t and connects ah and ah'. Note that this last property does not depend on G_0 being a normal subgroup.$\square$.

Definition: The quotient group

$$\pi_0(G)=: G/G_0 \tag{1.47}$$

is called th *zeroth homotopy group of G.*

In general, *all the order parameter spaces (the ones considered up to here and those to be considered in the following) possess a group of transformations acting* <u>*transitively*</u> *on the space itself.* We recall [30] that the action of G on T will be termed *transitive* iff, given any two points x,y $\in$ T, there exists (at least) one element of G whose action carries x into y.

Let then T be an order parameter space, and let G be the corresponding (Lie) group of transformations.

Remark: The choice of G may be not unique. For example, in the case of, say, He^4, with the order parameter (neglecting the constant ψ_0)

$$\psi(\vec{r})= \exp[i\phi(\vec{r})] \tag{1.48}$$

G can be chosen as

$$\exp[i\phi] \rightarrow \exp[i\phi] \cdot \exp[i\theta] \equiv \exp[i(\phi+\theta)] \tag{1.49}$$

leading to $G=U(1)$, or as

$$\phi \rightarrow \phi+\theta \tag{1.50}$$

which leads instead to $G= \mathbb{R}$. In the same way, for three-dimensional spins, one can consider $G=SO(3)$, but also $G=SU(2)$ (in the vector representation).

If G acts *effectively* on $\mathbb{T}$, i.e. if:

$$gx= x \Leftrightarrow g= e \;\; \forall \, x \in \mathbb{T} \tag{1.51}$$

then there is an obvious bijection between G and $\mathbb{T}$, which can be topologized in such a way that **$\mathbb{T}$ *becomes the group manifold of G*** . Hence, we can identify $\mathbb{T}$ with G itself. The identification proceeds as follows: take any $x_0 \in \mathbb{T}$ as a "reference" order parameter. Then any $x \in \mathbb{T}$ can be associated bijectively with the unique $g \in G$ such that $x=gx_0$. An example of this sort is that of planar spins, with $G=SO(2)$, but this kind of situation seldom happens in other cases. In general, instead, to each point $x \in \mathbb{T}$ there will be associated a subset of elements of G (it is left as an exercise to prove that it is actually a subgroup) leaving x invariant. We then define:

Definition: The *little group* (or the *isotropy subgroup*) of a point $x \in \mathbb{T}$

is the subgroup:

$$H_x = \{ \, g \in G \vdash gx = x \}$$

(1.52)

For example, for three-dimensional spins and $G = SO(3)$, one easily finds $H_x \sim SO(2) \; \forall x \in T = S^2$.

Theorem: The little groups at any two points $x, y \in T$ are *conjugate subgroups* in G.

Indeed, let g be such that $gx = y$. Then

$$h \in H_x \Leftrightarrow ghg^{-1} \in H_y$$

(1.53)

□

From now on, the little group of the reference order parameter (to which all the others are isomorphic) will be indicated simply by H. It will be independent from the choice of x_0 iff it is a normal subgroup, in which case G/H will itself be a group.

Note that, for any $x \in T$:

$$gx = g'x \Leftrightarrow g^{-1}g' \in H_x$$

(1.54)

Otherwise stated, g and gh, $h \in H_x$ transform x into the same point and viceversa. It is *the set of left cosets* of H_x that acts *effectively* (i.e. in a one-to-one way) to transform x into any other point of T. This is the basis of the following

Theorem: *The order parameter space* $\mathbb{T}$ *can be identified with the space of (left) cosets of* $\mathbb{H}$ *in* $\mathbb{G}$:

$$\mathbb{T} \simeq \mathbb{G}/\mathbb{H} \qquad (1.55)$$

Any such space is also called a <u>homogeneous space</u> for $\mathbb{G}$.☐

That the correspondence between $\mathbb{T}$ and $\mathbb{G}/\mathbb{H}$ is one-to-one has already been proved. We must prove that it is also continuous. Actually, this needs a definition of continuity (i.e. a topology) in the space of cosets, which has been constructed, up to now, in a purely algebraic manner. We then adopt the following [72]:

Definition: A sequence $\{\mathbb{H}_n\}$ of cosets is **convergent** iff (at least from some n onwards) it can be represented by a sequence $\{g_n\mathbb{H}\}$, with $\{g_n\}$ a convergent sequence in $\mathbb{G}$. It follows at once that to a convergent sequence of cosets there corresponds a convergent sequence of elements in $\mathbb{T}$. Conversely, if $x_n \rightarrow x_0$, we can choose the representation: $x_n = g_n x_0$ in such a way that, at least from some n onwards, g_n stays in an arbitrary neighborhood of the identity. Hence $\{g_n\}$ is convergent, and so is $\{g_n\mathbb{H}\}$. Convergence at any other point x can be analyzed by first "moving" x to x_0 with an appropriate group element, and then proceeding as before.☐.

EXAMPLES.

1) Planar spins. We have seen that one choice for G is $G=SO(2)$. In this case, $H=0$, and

$$T= SO(2) \tag{1.56}$$

We can however make other choices, namely:

1-a) $G= O(2)$. In this case, taking, e.g., x_0 as the spin aligned along the x-axis, the isotropy group of x_0 consists of the identity and the reflection in the y-axis. Hence:

$$H= \mathbb{Z}_2 , \quad \text{and:} \quad O(2)/\mathbb{Z}_2= SO(2) \tag{1.57}$$

(Note that $\mathbb{Z}_2$ is a normal subgroup of $O(2)$, and indeed: $O(2)=SO(2) \times \mathbb{Z}_2$).

1-b) $G=\mathbb{R}$. G acts then on the angle θ a spin makes with the reference direction of x_0 as:

$$\theta \rightarrow \theta+ 2\pi\alpha , \quad \alpha \in \mathbb{R} \tag{1.58}$$

It is then easy to see that

$$H=\mathbb{Z} \text{ (normal subgroup) ;} \quad \mathbb{R}/\mathbb{Z}= SO(2) \tag{1.59}$$

A digression on SO(3) and SU(2).

SU(2) is the group of unitary, unimodular 2×2 complex matrices. It can be parametrized as follows:

$$s \in SU(2) \Leftrightarrow s = \begin{bmatrix} \alpha & -\beta^* \\ \beta & \alpha^* \end{bmatrix} \; ; \; ss^\dagger = 1, \; \det(s) = 1 \Leftrightarrow |\alpha|^2 + |\beta|^2 = 1 \tag{1.60}$$

Therefore, the group manifold of SU(2) is the three-sphere S^3, and SU(2) is compact and simply connected.

As for any Lie group, SU(2) has also a "canonical" presentation [30,84] in terms of exponentials of the elements of the corresponding Lie algebra. In the representation (1.60) the Lie algebra of SU(2) can be identified with that of the Pauli matrices, and we can write:

$$s = \exp[i\vec{x} \cdot \vec{\sigma}] \tag{1.61}$$

with $\vec{x}$ a vector in $\mathbb{R}^3$. As:

$$\exp[i\vec{x} \cdot \sigma] \equiv \cos|\vec{x}| + i(\hat{x} \cdot \vec{\sigma})\sin|\vec{x}| \; ; \; \hat{x} = \frac{\vec{x}}{|\vec{x}|} \tag{1.62}$$

we find at once:

$$\alpha = \cos|\vec{x}| + i\hat{x}^3\sin|\vec{x}| \; ; \; \beta = i(\hat{x}^1 + i\hat{x}^2)\sin|\vec{x}| \tag{1.63}$$

Quite clearly, (1.63) is left invariant by the addition addition to $\vec{x}$ of any vector $\vec{a}$ parallel to $\vec{x}$ itself and such that: $|\vec{x}+\vec{a}| = |\vec{x}| + 2k\pi$, with k an integer. Hence, we can restrict $\vec{x}$ to the ball $\mathbb{B}_3$ of radius π. However, all the points on the two-sphere S^2 which is the boundary of $\mathbb{B}_3$ have to be identified, as they all lead to the same element of $SU(2)$, namely to the element parametrized by: $\alpha = -1$, $\beta = 0$, which is minus the identity. $\mathbb{B}_3$ with this identification becomes just S^3, as expected, just as, in a more easily visualizable example, the identification of all the points on the circumference of the two-ball (the disk) $\mathbb{B}_2$ reduces it to the two-sphere S^2.

There is a covering homomorphism $SU(2) \rightarrow SO(3)$, which can be constructed as follows.

Taking the Pauli matrices σ_i, i= 1,2,3, we can define:

$$\Sigma_i =: s\sigma_i s^{-1} =: \sigma_j \mathcal{R}_{ji}(s) \tag{1.64}$$

where the last equality (with summation over repeated indices) follows from Σ_i being hermitian and traceless, and from the σ_i's being a basis in the space of such matrices. One can prove by direct computation that

$$\Sigma_i^2 = \mathbb{1} \, , \, \text{i=1,2,3;} \tag{1.65-a}$$

and that

$$\Sigma_1 \Sigma_2 \Sigma_3 = i\mathbb{1} \tag{1.65-b}$$

$\mathbb{I}$ being the identity 2×2 matrix. Use of the identities

$$\sigma_j \sigma_k = i\epsilon_{jkl}\sigma_l + \delta_{jk}\mathbb{I} \tag{1.66-a}$$

$$\sigma_j \sigma_k \sigma_l = i\epsilon_{jkl}\mathbb{I} + \sigma_j \delta_{kl} - \sigma_k \delta_{jl} + \sigma_l \delta_{jk} \tag{1.66-b}$$

allows us to prove the following results:

i)
$$\Sigma_i^2 = \mathbb{I}\mathcal{R}_{ji}\mathcal{R}_{ij} + i\epsilon_{jkl}\sigma_l \mathcal{R}_{ji}\mathcal{R}_{ki} \tag{1.67}$$

(**No** summation over i). But the last term on the ´r.h.s. is zero by symmetry (the product of the $\mathcal{R}$'s being symmetric under the interchange j⇔k), and hence (1.62) implies:

$$(\tilde{\mathcal{R}}\mathcal{R})_{ii} = 1 \tag{1.68-a}$$

where $\tilde{\mathcal{R}}$ is the transpose matrix of $\mathcal{R}$ (it is superfluous to stress here that (1.64) implies that $\mathcal{R} = \| \mathcal{R}_{ij} \|$ is a 3×3 matrix with **real** entries).

ii) $\Sigma_1 \Sigma_2 \Sigma_3 = i\mathbb{I}(\det\mathcal{R}) + s[\sigma_1(\tilde{\mathcal{R}}\mathcal{R})_{23} - \sigma_2(\tilde{\mathcal{R}}\mathcal{R})_{13} + \sigma_3(\tilde{\mathcal{R}}\mathcal{R})_{12}]s^{-1}$

$$\tag{1.68-b}$$

which, together with (1.65-b), implies:

$$\det\mathcal{R} = 1; \quad (\tilde{\mathcal{R}}\mathcal{R})_{ij} = 0 \text{ for } i \neq j \tag{1.69}$$

All in all, we have proved that: $\tilde{\mathcal{R}}(s)\mathcal{R}(s)= \mathbb{1}$, and: $\det\mathcal{R}=1$, i.e. that $\mathcal{R}(s) \in SO(3)$.□

Note that, if $\mathcal{R}(s)=\mathcal{R}(s')$, then (1.64) implies

$$s^{-1}s'\sigma_i= \sigma_i s^{-1}s' \quad \forall i \tag{1.70}$$

and hence, by Schur's Lemma:

$$s^{-1}s'= \alpha\mathbb{1} \tag{1.71}$$

But: $\det(s^{-1}s')=1 \Rightarrow \alpha^2=1 \Rightarrow \alpha=\pm 1$, i.e. s and -s, and only them, determine a given element in $SO(3)$. Finally, note that

$$(ss')\sigma_i(ss')^{-1}= s(\sigma_j\mathcal{R}_{ji}(s'))s^{-1}= \sigma_k\mathcal{R}_{kj}(s)\mathcal{R}_{ji}(s')= \sigma_k[\mathcal{R}(s)\mathcal{R}(s')]_{ki} \tag{1.72}$$

This achieves the proof that the map $s\rightarrow\mathcal{R}(s)$ defined by (1.61) is a *two-to-one (covering) homomorphism of SU(2) onto SO(3)*.□

Exercise: Construct (using (1.64)) the lifting to $SU(2)$ of the $SO(2)$ subgroup of $SO(3)$ consisting of the rotations around the z-axis.

Rotation matrices in $SO(3)$ corresponding to rotations around the z-axis are of the form:

$$\mathcal{R} = \begin{vmatrix} \cos\theta & -\sin\theta & 0 \\ \sin\theta & \cos\theta & 0 \\ 0 & 0 & 1 \end{vmatrix} \qquad (1.73)$$

and one proves by direct computation that they lift to

$$s(\theta) = \pm \exp[-i\sigma_3 \tfrac{\theta}{2}] \in SU(2) \qquad (1.74)$$

(Note that: $s(\theta+2\pi) = -s(\theta)$, as $\exp[i\pi\sigma_3] = -1$). So, *the entire SO(2) lifts to a U(1) subgroup of SU(2) which covers it twice* (that's why $\theta/2$ appears in in (1.74) instead of θ). As $SU(2)$ is simply connected (the three-sphere S^3 is), it is also the *universal covering group* [78,90] of $SO(3)$.

The Hopf fibration $S^1 \Rightarrow S^3 \Rightarrow S^2$.

Remember that $S^1 \sim U(1)$, and $S^3 \sim SU(2)$, and consider, for $s \in SU(2)$, the quantity $s\sigma_3 s^{-1}$. It can be written clearly as

$$s\sigma_3 s^{-1} = \vec{x} \cdot \vec{\sigma}, \quad \vec{x} \in \mathbb{R}^3, \quad |\vec{x}| = 1$$

$$\qquad (1.75)$$

$$x^i = \tfrac{1}{2} \operatorname{Tr} [\sigma_i s\sigma_3 s^{-1}]$$

In terms of the parametrization (1.60) of $\mathsf{SU}(2)$, we also have:

$$x^1 + ix^2 = 2\alpha^*\beta \; ; \quad x^3 = |\alpha|^2 - |\beta|^2 \tag{1.75'}$$

Hence, (1.75) defines a **projection** from $\mathsf{S}^3 \sim \mathsf{SU}(2)$ onto S^2. If s and s' project down to the same $\vec{x} \in \mathsf{S}^2$, then h=: $s^{-1}s'$ commutes with σ_3 and, conversely, s and s·h will project down to the same $\vec{x}$ if h commutes with σ_3. With reference to (1.63), this implies $\beta=0$, i.e:

$$h = \exp[i\mu\sigma_3] \, , \quad \mu \in \mathbb{R} \tag{1.76}$$

But the set of such elements defines a $\mathsf{U}(1)$ subgroup of $\mathsf{SU}(2)$, and it follows then that the space of left cosets of $\mathsf{U}(1)$ is given by

$$\mathsf{SU}(2)/\mathsf{U}(1) = \mathsf{S}^2 \tag{1.77}$$

We then have a fiber bundle [30,94] with total space $\mathsf{S}^3 \sim \mathsf{SU}(2)$, fiber $\mathsf{S}^1 \sim \mathsf{U}(1)$ and base S^2. We will denote by: π: $\mathsf{SU}(2) \rightarrow \mathsf{SO}(3)$ the covering projection. The bundle is however **not** a trivial bundle. If it were, we would have: $\mathsf{S}^3 = \mathsf{S}^2 \times \mathsf{S}^1$. But then it would follow: $\pi_1(\mathsf{S}^3) = \pi_1(\mathsf{S}^2) \times \pi_1(\mathsf{S}^1)$, which is not possible, as $\pi_1(\mathsf{S}^n)=0$ for $n \geq 2$, while $\pi_1(\mathsf{S}^1)=\mathbb{Z}$.

Remark. We have employed here implicitly the following

Theorem: If $\mathfrak{X}$ and $\mathfrak{Y}$ are topological spaces, then:

$$\pi_1(\mathcal{X} \times \mathcal{Y}) = \pi_1(\mathcal{X}) \times \pi_1(\mathcal{Y}) \tag{1.78}$$

The proof of the theorem is simple, and will be omitted here.□.

After this long digression, let's turn to another (and last) example, namely to

2) Ordinary Spins. It is quite obvious that we can take $G=SO(3)$, and that the little group of any $\vec{x} \in S^2$ is just $SO(2)$. One easily proves that:

$$SO(3)/SO(2) = S^2 \tag{1.79}$$

(indeed, in general: $SO(n+1)/SO(n) = S^n$). Note also that, according to (1.77):

$$SO(3)/SO(2) \sim SU(2)/U(1) \tag{1.80}$$

i.e. that the quotient is **unchanged** by lifting both groups on the l.h.s. into $SU(2)$ (to be more precise, we have pulled back both $SO(3)$ and $SO(2)$ to $SU(2)$ via the covering projection).

1.7. PRELIMINARY THEOREMS CONCERNING FUNDAMENTAL GROUPS.

We start by considering the (rather exceptional) case in which the order parameter space is itself a group ($\mathbb{H}=0$). Then:

Theorem: If G is a Lie group, then

$$\pi_1(G) \text{ is } \textbf{\textit{Abelian.}} \tag{1.81}$$

In view of the isomorphism between based homotopy groups, it will be sufficient to compute the based homotopy group $\pi_1(G,e)$, with e the identity in G. To prove the result, consider the map:

$$\Phi = [0,1] \times [0,1] \rightarrow G \quad \text{by: } (u,v) \rightarrow f(u) \cdot g(v) \tag{1.82}$$

with f and g two loops at e, i.e.:

$$f(0)=f(1)=g(0)=g(1)=e \tag{1.83}$$

and $f(u) \cdot g(v)$ being given by the group multiplication.

For any path C in the unit square $I \times I$ beginning at $(0,0)$ and ending at the opposite corner $(1,1)$, the restriction $\Phi|_C$ of the map (1.82) to the path C defines clearly a loop in G at e. If two paths, such as those labeled (a) and (b) in Fig. 9, are homotopic in the square (with endpoints fixed), the corresponding loops in G will be also homotopic,

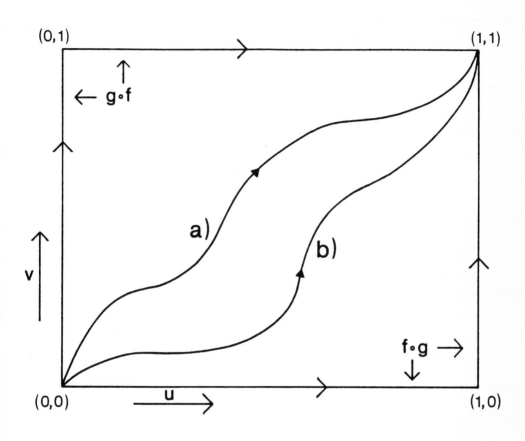

Fig. 9. a) and b): Generic homotopic paths in $[0,1] \times [0,1]$ connecting the points $(0,0)$ and $(1,1)$. The loops $f \cdot g$ and $g \cdot f$ correspond to opposite pairs of subsequent sides of the square.

the homotopy being provided by that needed to deform the two paths in the square into each other. In particular, the two paths joining the origin to the opposite corner and running along the edges of the square will correspond (after an essentially trivial reparametrization) to $f \cdot g$ (lower one) and to $g \cdot f$ (upper one) respectively. As the two paths are homotopic in the square, so will be $f \cdot g$ and $g \cdot f$. This achieves the proof of the theorem.$\square$

We discuss now a more general theorem concerning homogeneous spaces, namely:

Theorem: Let G be a ***connected and simply connected*** Lie group, let H be any closed subgroup of G, and let H_0 be the connected component of the identity of H. Then:

$$\pi_1(G/H) = \pi_0(H) = H/H_0 \tag{1.84}$$

The proof of this theorem will be supplied in a later Section. Let's discuss for the time being some of its consequences on various examples:

i) **Planar Spins:** $G = T(1)$, $H = \mathbb{Z} \Rightarrow H_0 = 0$, and $\pi_1(G/H) = \mathbb{Z}$.

ii) **Ordinary Spins:** $G = SU(2)$, $H = U(1)$ (the proper lift of $SO(2)$ into $SU(2)$). H is connected, and hence $\pi_0(H) = \pi_1(G/H) = 0 = \pi_1(S^2)$.

iii)**Nematics:** $G=SU(2)$, but now $H=U(1) \times \mathbb{Z}_2$, and hence $\pi_1(G/H)=$
$\mathbb{Z}_2 = \pi_1(\mathbb{RP}^2)$.

iv)**Biaxial Nematics:** While the order parameter of nematics is invariant
under the full group of rotations in a plane plus a π rotation around an
axis orthogonal to the plane (corresponding to the fact that the order
parameter is associated with a direction in ordinary space), *biaxial*
nematics are characterized by an order parameter which is invariant
only under discrete symmetries, namely rotations of π around three
mutually orthogonal axes. These systems are characterized therefore
[72] by $G=SO(3)$ and $H=D_2$, the *dihedral group* : $D_2= \{\mathbb{I},\ \mathbb{P}_x,\mathbb{P}_y.\ \mathbb{P}_z\}$,
where $\mathbb{P}_i$, i=x,y,z is a rotation of an angle π around the i-th axis. D_2
lifts into $SU(2)$ to the *group of quaternions* **Q** defined by:

$$Q= \left\{\pm \mathbb{I},\ \pm i\sigma_x,\ \pm i\sigma_y,\ \pm i\sigma_z\right\} \qquad (1.85)$$

Then: $SO(3)/D_2 \sim SU(2)/Q$ and, as **Q** is discrete (the connected
component of the identity consists of the identity alone):

$$\pi_1(SU(2)/Q)= Q \qquad (1.86)$$

This is the first explicit example we encounter of a medium with a
nonabelian fundamental group (the previous example of the figure-eight
cannot be associated with any physical distribution of order
parameters).

1.8. HIGHER HOMOTOPY GROUPS.

The Second Homotopy Group $\pi_2(\mathsf{T})$.

As in the discussion of π_1, let us start with the discussion of *based* homotopies. We consider then *maps of the two-sphere S^2 into T which send a given point of S^2 into a fixed point $x \in \mathsf{T}$* or, equivalently, maps of the form (see Fig. 10-a):

$$f\colon I \times I \to \mathsf{T}; \quad I = [0,1]; \quad (u,v) \to f(u,v) \in \mathsf{T} \tag{1.87}$$

s.t.:

$$f(0,v)=f(1,v)=f(u,0)=f(u,1)=x \in \mathsf{T}, \text{ or, in short: } f(\partial(I \times I)) \equiv x \tag{1.88}$$

A **homotopy** between any two such maps, f and g, will be defined as a map:

$$H\colon I \times I \times I \to \mathsf{T}; \quad H_z\colon I \times I \to \mathsf{T}, z \in I \tag{1.89}$$

s.t.:

$$H_z(\partial(I \times I)) \equiv x \quad \forall z$$
$$H_0 = f \tag{1.90}$$
$$H_1 = g$$

The **product** of two maps f and g will be defined as:

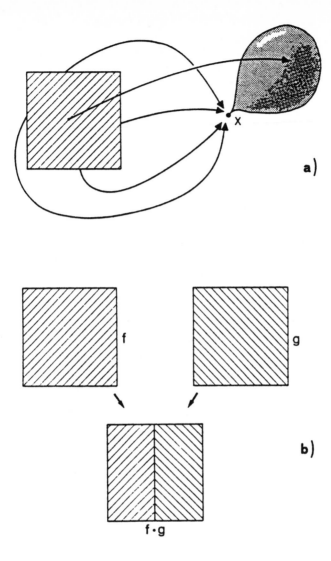

Fig. 10. a) The map definimg a two-loop in $\mathbb{T}$ at x. b) Pictorial representation of the product of a pair of two-loops, Eq. (1.91).

$$(f \cdot g)(u,v) = \begin{cases} f(2u,v), & 0 \le u \le 1/2 \\ \\ g(2u-1,v), & 1/2 \le u \le 1 \end{cases} \qquad (1.91)$$

and is depicted in Fig. 10-b.

Theorem: $f \cdot g \sim g \cdot f$. The homotopy is depicted in Fig. 11. $\square$.

By the same technique one can prove that:

$$f \sim f', \, g \sim g' \Rightarrow f \cdot g \sim f' \cdot g' \qquad (1.92\text{-a})$$

$$(f \cdot g) \cdot h \sim f \cdot (g \cdot h) \qquad (1.92\text{-b})$$

$$e \cdot f \sim f \cdot e \sim f \ \ \forall \, f \qquad (1.92\text{-c})$$

$$f \sim f' \Rightarrow f^{-1} \sim f'^{-1} \qquad (1.92\text{-d})$$

$$f \cdot f^{-1} \sim f^{-1} \cdot f \sim e \qquad (1.92\text{-e})$$

where we define:

$$f^{-1}(u,v) = f(1-u,v)$$

$$(1.93)$$

$$e(u,v) \equiv x \ \ \forall \, u,v$$

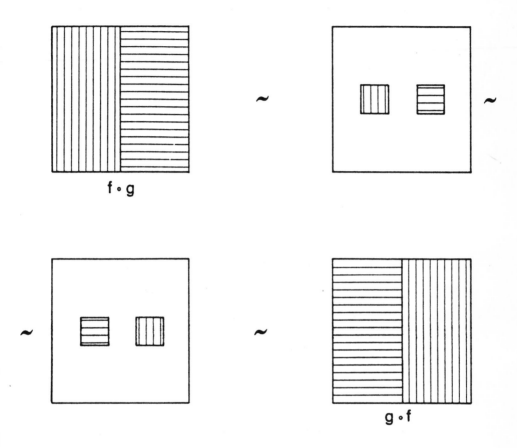

Fig. 11. Homotopic commutativity of the product of two-loops. Shrinking f and g to the small inscribed squares (first step) with the white area all mapped into the single point x, one can move the small squares around (second step), and then "enlarge" them back to the original size, thereby obtaining the final result.

Let us sketch the proof of (1.92-e), which is the only property which is not almost self-evident. Let's represent [72] the image of α as a sphere in $\mathbb{T}$ with the "u-lines" (v= const.) corresponding to the "meridians" of the sphere, the "v-lines" (u=const.) to the "parallels". Let us also recall that:

$$(f \cdot f^{-1})(u,v) = \begin{cases} f(2u,v) & 0 \leq u \leq 1/2 \\ \\ f(2(1-u),v) & 1/2 \leq u \leq 1 \end{cases} \qquad (1.94)$$

Hence, each "meridian" is traced back and forth and, proceeding as in the case of π_1, we can "open up a hole" in the sphere and subsequently retract the "bag" thus obtained to a single point (Fig. 12). $\square$.

We then have the following:

Theorem:

The set $\{[f]\}$ of homotopy classes of two-loops based at $x \in \mathbb{T}$ has a group structure, the group operations being defined as:

$$[f] \cdot [g] = [f \cdot g] \qquad (1.95\text{-a})$$
$$[e] \cdot [f] = [f] \cdot [e] = [f] \qquad (1.95\text{-b})$$
$$[f]^{-1} = [f^{-1}] \qquad (1.95\text{-c})$$

$\square$

Definition: The group defined in the previous theorem is the *second honotopy group based at x*, and will be denoted by $\pi_2(\mathbb{T},x)$.

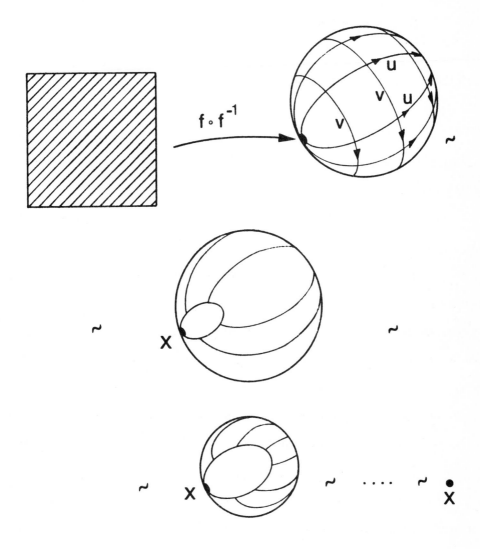

Fig.12. Illustration of the procedure leading to the proof of (1.92-e).

Theorem: $\pi_2(\mathbb{T},x)$ is *abelian* $\forall\, x \in \mathbb{T}$.

This follows from $f \cdot g \sim g \cdot f$. $\square$

Theorem: $\pi_2(\mathbb{T},x) \sim \pi_2(\mathbb{T},y)$ $\forall\, x,y \in \mathbb{T}$.

The explicit construction proceeds as in the case of π_1, and is depicted in Fig. 13, where we add to f an "umbilical cord" provided by a path c connecting y to x, which turns it into a two-loop $c \cdot f$ at y. The cord may be viewed as a degenerate part of the loop $c \cdot f$. The procedure works also in the opposite direction: if g is a two- loop at y, $c^{-1} \cdot g$ will be a two-loop at x. In order to keep the parametrization inside the unit

square, we must however "shrink" f (or g) to a smaller square inscribed inside the original unit square. The procedure is illustrated in Fig. 13-b for $c \cdot f$, where f has been shrunk to the small shaded square in the center of the unit square, and each one of the outer concentric square contours is mapped into a single point of c, c(z), until eventually the border $\partial(I \times I)$ is mapped into c(0)=y.

We show now explicitly how this "shrinking" procedure can be performed in a homotopic way, by building the homotopy which shrinks a map defined on $I \times I$ to a map defined on, say, $[\tau,1\text{-}\tau] \times [\tau,1\text{-}\tau]$, $0 \le \tau < 1/2$, i.e. to a smaller square concentrical to the original one.

Consider the map:

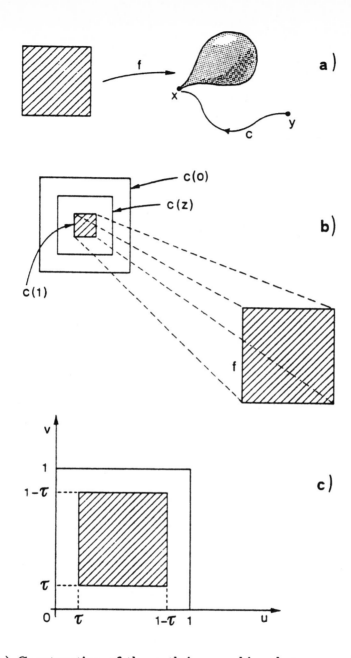

Fig. 13. a) Construction of the path-isomorphism between $\pi_2(\mathbb{T},x)$ and $\pi_2(\mathbb{T},y)$. b) A more formal construction of $c([f])$, obtained by shrinking f to the small shaded square, while mapping the boundaries of each inscribed square into $c(z)$, $0 \le z \le 1$. c) Illustratrion of Eq. (1.96).

$$\tilde{f}_\tau = \begin{cases} f(0,0) \text{ in the white area of Fig. (13-c)} \\ \\ f(\frac{u-\tau}{1-2\tau}, \frac{v-\tau}{1-2\tau}) \text{ in the shaded square} \end{cases} \qquad (1.96)$$

Clearly:

$$h_t(u,v) =: \tilde{f}_{t\tau}(u,v) \qquad (1.97)$$

is the required homotopy between f and $\tilde{f}_\tau$, which is the "shrunk" version of f.

Returning now to the proof of the theorem, it is easy to show that the map

$$c: \pi_2(\mathbb{T},x) \rightarrow \pi_2(\mathbb{T},y) \qquad (1.98)$$

depicted in Fig. 13-a) does respect homotopies, i.e. $f \sim g \Rightarrow c(f) \sim c(g)$, and also that, if c and c' are two homotopic paths (i.e. $[c \cdot c'^{-1}]$ is the identity in $\pi_1(\mathbb{T},y)$), then $c(f) \sim c'(f)$, i.e that the (path) isomorphism between based second homotopy groups is **canonical**.☐.

We conclude that there is a **unique abstract group,** $\pi_2(\mathbb{T})$, the **second homotopy group of** **T**, of which the based second homotopy groups are isomorphic realizations, and that $\pi_2(\mathbb{T})$ is an **abelian** group.

Higher Homotopy Groups.

All that has been said concerning two-loops can be transferred without changes to **n-loops**, defined as maps:

$$f: I \times I \times \ldots \times I \rightarrow \mathbb{T} \; ; \quad f(\partial(I \times \ldots \times I)) = x \in \mathbb{T} \qquad (1.99)$$

and we can proceed as before to the construction of the (abelian) **n-th homotopy group based at x**, $\pi_n(\mathbb{T},x)$, and of the abstract **n-th homotopy group**, $\pi_n(\mathbb{T})$. We will not insist in details here.

Calculation of π_2 for homogeneous spaces.

Theorem: If G is a Lie group, $\mathbb{H}$ a closed subgroup of G, and

$$\pi_1(G) = \pi_2(G) = 0 \qquad (1.100)$$

Then:

$$\pi_2(G/\mathbb{H}) \sim \pi_1(\mathbb{H}_0) \qquad (1.101)$$

where $\mathbb{H}_0$ is the connected component of the identity of $\mathbb{H}$.

We will omit the proof for the time being, as it will be provided in a later Section. $\square$.

Let us also recall that, according to a theorem due to E. Cartan (for whose proof the reader is referred to the literature [57]):

$$\pi_2(G) = 0 \text{ for } \mathbf{compact} \text{ Lie groups} \qquad (1.102)$$

$$\square$$

We now discuss some applications of the previous theorems:

i) Planar Spins: $T = S^1$ can be identified with $\mathbb{R}/\mathbb{Z}$. As $\mathbb{Z}(\equiv H)$ is discrete, and $\pi_2(\mathbb{R}) = 0$:

$$\pi_2(S^1) = o \qquad (1.103)$$

ii) Ordinary Spins: We already know that $T = S^2 \sim SU(2)/U(1)$. In this case $H \equiv H_0 = U(1)$, and hence:

$$\pi_2(S^2) \sim \pi_1(U(1)) = \mathbb{Z} \qquad (1.104)$$

iii) Nematics: $T = \mathbb{R}P^2 = S^2/\mathbb{Z}_2$. Lifting to $SU(2)$, the isotropy subgroup H

may be identified with the two-component group composed of $U(1)$ and $U(1) \cdot i\sigma_y$. Then:

$$\pi_2(\mathbb{R}P^2) = \pi_1(U(1)) = \mathbb{Z} \qquad (1.105)$$

iv) Biaxial Nematics: $T = SU(2)/Q$, where Q is the quaternion group As Q is a discrete group:

$$\pi_2(SU(2)/Q) = 0 \qquad (1.106)$$

1.9. RELATIVE HOMOTOPY AND RELATIVE HOMOTOPY

GROUPS.

We will consider here essentially only second homotopy groups, i.e n=2. Generalizations of what will be said to higher n's are straightforward.

Relative homotopy groups are defined w.r.t. a base-point x *and* a subset A of T, with $x \in A$, by introducing "loops" which send three faces (edges, for n=2) of the square $I \times I$ into x and the fourth face into A, while the interior goes into T as before. Previous "absolute" homotopy groups are reobtained for $A \equiv x$ (i.e. when A shrinks to a point). n-th order "relative loops" for $n > 2$ are defined similarly. Also, *the loop product is defined exactly as before*, and we are thus led in a natural way to define *relative (based) homotopy groups*. The group based at x will be denoted by $\pi_2(T,A,x)$.

Remark: The proof that π_2 is abelian cannot be repeated here, and **$\pi_2(T,A,x)$ need not be Abelian.** Indeed, if f is a "relative loop", f(0,v)=f(1,v)=f(u,1)=x, while we can only assert that $f(u,0) \in A$. In other words, **f(u,0) is a one-loop contained in A (and based at x)**. But then:

$$(f \cdot g)(u,0) = \begin{cases} f(2u,0) & 0 \le u \le 1/2 \\ \\ g(2u\text{-}1,0) & 1/2 \le u \le 1 \end{cases} \tag{1.107}$$

while in $(g \cdot f)(u,0)$ the loops in $\mathbb{A}$ associated with f and g appear in the reverse order. Therefore:

Theorem: $\pi_2(\mathbb{T},\mathbb{A},x)$ will be Abelian iff $\pi_1(\mathbb{A})$ is. $\square$

However, it can be proved [57] (we will not do it here) that *$\pi_n(\mathbb{T},\mathbb{A},x)$ is Abelian for $n \geq 3$.* $\square$

1.10. THE EXACT HOMOTOPY SEQUENCE.

Consider the groups $\pi_n(\mathbb{A},x)$, $\pi_n(\mathbb{T},x)$, $\pi_n(\mathbb{T}.\mathbb{A},x)$ and $\pi_{n-1}(\mathbb{A},x)$. There are various maps that can be established among them, all of which are actually **homomorphisms**, namely:

i) As $\mathbb{A} \subset \mathbb{T}$, a map of a cube into $\mathbb{A}$ at x is also a map of the cube into $\mathbb{T}$ at x, whence the homomorphism:

$$\alpha_n : \pi_n(\mathbb{A},x) \rightarrow \pi_n(\mathbb{T},x) \qquad (1.108)$$

(essentially an identification mapping).

ii) As $x \in \mathbb{A}$, a map of a square into $\mathbb{T}$ at x is a special case of a map into $(\mathbb{A},x)$ (when the loop at x in $\mathbb{A}$ corresponding to one of the faces of the cube reduces to the trivial loop). We then have the homomorphism:

$$\beta_n \colon \pi_n(\mathbb{T}, x) \to \pi_n(\mathbb{T}, \mathbb{A}, x) \tag{1.109}$$

For the case n=2, this is illustrated in Fig. 14-a. Note that the identity in $\pi_n(\mathbb{T}, \mathbb{A}, x)$ is the homotopy class of maps whose image can be continuously deformed into an image lying entirely in $\mathbb{A}$. But this is both the kernel of β_n and the image of α_n. Hence, we have the following:

Theorem: $\qquad\qquad \mathrm{Im}(\alpha_n) = \mathrm{Ker}(\beta_n).$ $\Box.$ $\qquad\qquad$ (1.110)

Next, observe that any map of an n-cube into $\mathbb{T}$ with one face (an (n-1)-cube) mapped into $\mathbb{A}$ is also a map of an (n-1)-cube into $\mathbb{A}$ at x, whence the homomorphism:

$$\gamma_n \colon \pi_n(\mathbb{T}, \mathbb{A}, x) \to \pi_{n-1}(\mathbb{A}, x) \tag{1.111}$$

The image of β_n is the set of loops whose "loop in $\mathbb{A}$" is homotopic to the constant loop, i.e. it is precisely the kernel of γ_n.

Hence (see Fig. 14-b):

Theorem: $\qquad\qquad \mathrm{Im}(\beta_n) = \mathrm{Ker}(\gamma_n).$ $\Box.$ $\qquad\qquad$ (1.112)

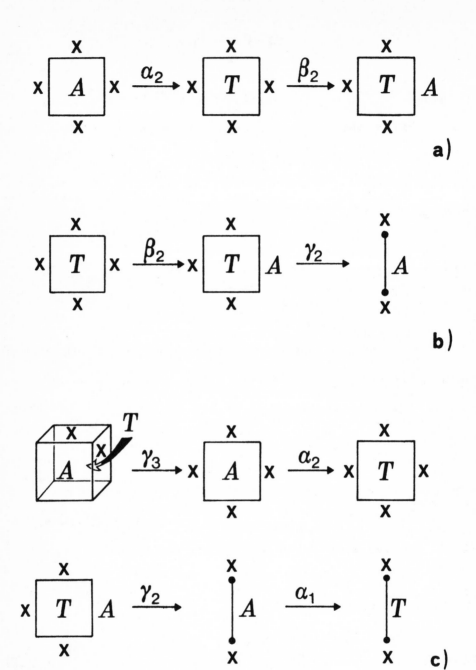

Fig. 14. a) An illustration of Eq. (1.118) for n=2. b) Same for Eq.(1.120). c) Same for Eq. (1.121), n=3 and n=2.

Finally, one can also prove (see Fig. 14-c) that:

Theorem: $\qquad\qquad\qquad \mathrm{Im}(\gamma_n) = \mathrm{Ker}(\alpha_{n-1}). \;\square \qquad\qquad\qquad$ (1.113)

Indeed, the first cube in Fig. 14-c) can be seen as the unfolding, *in* T, of a homotopy shrinking the "A" face into the single point x, the homotopy being provided by the coordinate normal to the face. Hence, all (n-1)-loops in A provided by γ_n are actually homotopic to the trivial loop when viewed, via α_{n-1}, as loops in T. $\square$.

We recall [94] that whenever a sequence of group homomorphisms is such that the image of any one of them coincides with the kernel of the successive one, while its kernel is in turn the image of the previous one, the sequence is said to be **exact**. Therefore, what we have eventually succeeded is in establishing is the following, fundamental

Theorem: The *long homotopy sequence*:

$$... \to \pi_n(A,x) \xrightarrow{\alpha_n} \pi_n(T,x) \xrightarrow{\beta_n} \pi_n(T,A,x) \xrightarrow{\gamma_n}$$

$$\to \pi_{n-1}(A,x) \xrightarrow{\alpha_{n-1}} \pi_{n-1}(T,x) \xrightarrow{\beta_{n-1}} \pi_{n-1}(T,A,x) \xrightarrow{\gamma_{n-1}}$$

$$\to \dots\dots\dots\dots\dots\dots\dots\dots\to \qquad\qquad (1.114)$$

$$\to \pi_1(A,x) \xrightarrow{\alpha_1} \pi_1(T,x) \xrightarrow{\beta_1} \pi_1(T,A,x) \xrightarrow{\gamma_1}$$

$$\to \pi_0(A,x) \xrightarrow{\alpha_0} \pi_0(T)$$

is an __exact__ sequence.□.

In the sequel, we will need also the following:

Theorem: If the (short) sequence of group homomorphisms:

$$0 \rightarrow G_1 \overset{\phi}{\rightarrow} G_2 \rightarrow 0 \qquad (1.115)$$

is exact, then ϕ: $G_1 \rightarrow G_2$ is an **isomorphism**.

Indeed, exactness of the sequence implies:

i) $\mathrm{Im}(0 \rightarrow G_1) = \mathrm{Ker}(\phi) \Rightarrow \mathrm{Ker}(\phi) = \{$Identity element of $G_1)$. Therefore, ϕ is a one-to-one map.

ii) $\mathrm{Im}(\phi) = \mathrm{Ker}(G_2 \rightarrow 0) \Rightarrow \mathrm{Im}(\phi) = G_2$, i.e. ϕ is onto. □.

Applications of the exact sequence,

From now on we will identify T with a Lie group G, A with a subgroup $\mathsf{H} \subset \mathsf{G}$ and x with the identity e in G. Explicit mention of the latter will be omitted from now on. The main result we will employ here is that $\pi_n(\mathsf{G}/\mathsf{H})$ *can be identified with with the n-th relative homotopy group* $\pi_n(\mathsf{G},\mathsf{H},e)$, or $\pi_n(\mathsf{G},\mathsf{H})$ for short. For details of the proof, see [54,72,94]. The homotopy sequence becomes then:

$$\ldots \to \pi_n(\mathbb{H}) \to \pi_n(\mathsf{G}) \to \pi_n(\mathsf{G}/\mathbb{H}) \to \pi_{n-1}(\mathbb{H}) \to \ldots. \qquad (1.116)$$

Suppose now that $\pi_0(\mathsf{G})= \pi_1(\mathsf{G})=0$. From (1.116) we can then extraxt the *short exact sequence* :

$$0 \to \pi_1(\mathsf{G}/\mathbb{H}) \to \pi_0(\mathbb{H}) \to 0 \qquad (1.117)$$

proving that $\pi_1(\mathsf{G}/\mathbb{H}) \sim \pi_0(\mathbb{H})$.

Similarly, if $\pi_2(\mathsf{G})= \pi_1(\mathsf{G})= 0$, we have the short exact sequence:

$$0 \to \pi_2(\mathsf{G}/\mathbb{H}) \to \pi_1(\mathbb{H}) \to 0 \qquad (1.118)$$

proving that $\pi_2(\mathsf{G}) \sim \pi_1(\mathbb{H})$. We have thus proved two theorems that had been stated previously without proof. More generally, with a similar short sequence, one proves easily that:

$$\pi_n(\mathsf{G})=\pi_{n-1}(\mathsf{G})= 0 \Rightarrow \pi_n(\mathsf{G}/\mathbb{H}) \sim \pi_{n-1}(\mathbb{H}) \qquad (1.119)$$

Another result that we will prove in a short while is the following:

Theorem: $$\pi_n(S)^1 = 0 \; \forall n > 1 \tag{1.120}$$

□

Armed with this, and using the Hopf fibration $S^1 \to S^3 \to S^2$, we arrive at:

$$0 = \pi_3(S^1) \to \pi_3(S^3) \to \pi_3(S^2) \to \pi_2(S^1) = 0 \tag{1.121}$$

whence:

$$\pi_3(S^3) \sim \pi_3(S^2) \tag{1.122}$$

Later on we will prove that $\pi_n(S^n) = \mathbb{Z} \; \forall n$. Therefore:

$$\pi_3(S^2) \sim \mathbb{Z} \tag{1.123}$$

We end up by quoting the following theorem of Freudenthal [54]:

Theorem: $$\pi_k(S^n) \sim \pi_{k+1}(S^{n+1}) \text{ for } k < 2n - 1 \tag{1.124}$$

□

This has the following consequences:

i) As $\pi_1(S^n) = 0$ for $n > 1$:

$$\pi_k(S^n) \sim \pi_{k-1}(S^{n-1}) \sim \ldots \sim \pi_1(S^{n-k+1}) = 0 \quad \text{whenever } n-k+1 > 1$$

$$\tag{1.125}$$

Hence:

$$\pi_k(S^n) = 0 \text{ for } k < n \tag{1.126}$$

□

ii)
$$\pi_n(S^n) \sim \pi_2(S^2) \sim \mathbb{Z} \tag{1.127}$$

□

Finally, let's consider $G=\mathbb{R}$ and $H=\mathbb{Z}$. Then: $G/H \sim S^1$. Eq. (1.116) becomes:

$$\ldots \to \pi_n(\mathbb{R}) \to \pi_n(S^1) \to \pi_{n-1}(\mathbb{Z}) \to \pi_{n-1}(\mathbb{R}) \to \ldots \tag{1.128}$$

But $\pi_n(\mathbb{R})=0 \ \forall n$, $\pi_n(\mathbb{Z})=0 \ \forall n \geq 1$, while $\pi_0(\mathbb{Z})=\mathbb{Z}$. So, for n=1, we obtain:

$$0 \to \pi_1(S^1) \to \pi_0(\mathbb{Z}) \to 0 \tag{1.129}$$

proving that $\pi_1(S^1)=\mathbb{Z}$, a result we know already. For $n > 1$, we obtain:

$$0 = \pi_n(\mathbb{R}) \to \pi_n(S^1) \to \pi_{n-1}(\mathbb{Z}) = 0 \tag{1.130}$$

i.e.: $0 \to \pi_n(S^1) \to 0$, proving that

$$\pi_n(S^1) = 0 \ \forall \ n > 1 \qquad (1.131)$$

which is precisely Eq. (1.120).

CH.2. TOPOLOGICAL METHODS IN CLASSICAL FIELD THEORY.

2.1. INTRODUCTION.

After having introduced in Chapt.1 some of the basic technologies of homotopy theory, using the problem of defects in ordered media in order to visualize most of the concepts and theorems involved, we turn now to the study of some models of **classical** field theories.

A *classical field* [15,40,42,84]is a continuous map:

$$\Phi: \mathbb{M} \ni x \mapsto \vec{\phi}(x) \in \mathbb{T} \tag{2.1}$$

from a (spacetime) manifold $\mathbb{M}$ to a topological space $\mathbb{T}$, the field space. Alternatively, adopting a "nonrelativistic" point of view, a classical field can be viewed a family of maps from a space manifold $\mathbb{M}_0$ to the field space indexed by one real (time) variable. Although more general settings are possible, we will assume here $\mathbb{M}$ to be a flat (Minkowskian or Euclidean) manifold. Typically: $\mathbb{M}=\mathbb{R}^{d+1}$ for some integer d (the space dimension). In turn, $\mathbb{T}$ will be assumed to be a compact topological space. More specifically, $\mathbb{T}$ will be either a compact Lie group or, more generally, a coset space.

The similarity of what we are beginning to discuss here and the content of Ch. 1 should be evident. To some extent, the theory of defects in ordered media can be viewed as the study of the topological properties of *static*, singular field configurations. As such, the fields considered in that context have no dynamics. In the present context, we will be interested instead in studying *smooth* field configurations and their topological properties, in the static as well as in the dynamic case. The field dynamics will be assumed to be governed by an *action principle*, i.e. by evolution equations obtained by varying an action $\mathbb{S}$ which is a functional of the field values: $\mathbb{S} = \mathbb{S}[\vec{\phi}]$. The action is usually taken as the integral of a local *Lagrangian density*, i.e. of the form:

$$\mathbb{S}[\vec{\phi}] = \int_{\mathbb{M}} dx \mathcal{L}(x) \qquad (2.2)$$

where locality means that $\mathcal{L}$ is assumed to be a function of the field, of its derivatives up to a finite order and possibly of the spacetime coordinates. We will consider here only Lagrangians which contain derivatives up to first order. As a consequence, the action principle [84]:

$$\delta \mathbb{S} = 0 \; ; \; \delta\vec{\phi} = 0 \text{ at spacetime infinity} \qquad (2.3)$$

will yield field equations which will be second order PDE's.

One can define the total energy associated with a given field configuration either directly within the Lagrangian formalism or by going, in standard way, to the Hamiltonian formalism via the Legendre map [1,84]. Only field configurations which yield a finite total energy will be considered as physically acceptable. In the absence of gauge degrees of freedom (see below) , this requires that the fields go to some constant value at spacetime infinity. Under this situation, the spacetime manifold $M=\mathbb{R}^{d+1}$ can be considered as being actually **compactified** (via one-point compactification [90]) to the (d+1)-sphere S^{d+1}. Therefore, the field configurations will be classified in a natural way by the homotopy group $\pi_{d+1}(\mathbb{T})$ or, in the case of static field configurations (i.e. of static solutions of the field equations), by $\pi_d(\mathbb{T})$.

Warning. Although the one depicted above will be the "generic" situation that we shall consider in what follows, there are at least two important situations in which compactification of spacetime to a sphere is not at all natural, namely:

i) In (Quantum) Statistical Mechanics the rôle of time is taken by an imaginary variable running from 0 to $+i\hbar\beta$, where $\beta=(kT)^{-1}$, k is Boltzmann's constant and T is the temperature. The relevant correlation functions of the fields will obey periodic (for Bosons) or antiperiodic (for Fermions) boundary conditions along the "time" axis (Kubo-Martin-Schwinger (KMS) conditions), i.e. *"time" will be independently compactified to a circle S^1 of circumference $\hbar\beta$*. The natural compactification of the space-"time" manifold is therefore not to S^{d+1} but instead to $S^d \times S^1$, and the corresponding homotopies of maps from $S^d \times S^1$ to $\mathbb{T}$ can be quite different from the previous ones [80].

ii) Quite often, in problems in Condensed-Matter Physics, one considers (either directly or as the continuum limit of some discrete model) nonrelativistic field theories in which the fields obey periodic boundary conditions in space. In such a case, *space is independently compactified to a d-torus* $\mathcal{T}^d$, and the relevant homotopies become those of the maps from $\mathcal{T}^d \times \mathbb{R}$ (or from $\mathcal{T}^d \times S^1$ in Statistical Mechanics) to the field space $\mathbb{T}$. In this case as well the homotopies can be markedly different [49] from the case when the whole of spacetime is compactified to a sphere.

As already stated, we will stick here to the more pedagogic and "generic" case of the one-point compactification discussed above, having however in mind that, in specific situations, one may be forced to construct different homotopies.

Assume for the moment that the space of all field configurations can be decomposed into sectors classified by some homotopy group. At this level, it is actually not relevant which homotopy classifies the field configurations, provided there is some. It is quite clear that :

i) The different sectors are disjoint, i.e. the space of field configurations (which is an infinite-dimensional function space and *not* the space $\mathbb{T}$ of the field *values* !) falls into the disjoint union of the different homotopy sectors, and that:

ii) Small variations of the field configurations, such as those involved in the derivation of the field equations from the action principle (2.3), cannot lead from a given homotopy sector to a different one. In other words, the field equations are insensitive to which homotopy sector we are in (this latter information will be rather encoded into the boundary conditions which have to supplement the field equations). Also, time evolution, being by assumption a smooth process, cannot "lead outside" of any preassigned homotopy sector.

Suppose then that we add to the action a **topological term**, i.e. some functional of the field which depends only on the homotopy sector the field is in. We will see some specific examples of topological terms already in the next Section. It is quite clear that any such term will be insensitive to smooth (and, in particular, to small) modifications of the fields, and therefore the action modified in this way will lead to the same field equations. It is to be expected (and will be shown on examples in the sequel) that any such term can be represented by the addition of a total divergence to the Lagrangian. Therefore, we can conclude that the addition of topological terms to the action is completely harmless at the purely **classical** level. It can have however profound consequences at the **quantum** level. The easiest way of seeing this is to remember that, when quantizing a field theory with the Feynman path-integral procedure [40,41,65,87], each field configuration must be weighted by the "Feynman factor" $\exp[iS[\vec{\phi}]/\hbar]$, and the addition of topological terms to the action will lead to quantum interference effects between the contributions to the path-integral coming from different homotopy sectors. Such effects will be altogether absent at the classical level, where only the stationary "points" of the action functional (in the space of field configurations) will matter.

Topological field theories (i.e. field theories containing topological terms in the action or even theories with an entirely topological action) have been proposed by various authors and in different contexts [2,15,33,34,37,42,48,49,52,59,61-64,84,93,103,108,111,112,115,116]. In this Chapter, I will discuss some simple instances of such model field theories, and namely the (1+1) and (2+1) $SO(3)$ nonlinear σ-models [115]. The first one arises in a natural way as the continuum limit of a one-dimensional, antiferromagnetic Heisenberg chain. Affleck [2] and Haldane [52] have shown how, in the continuum limit, the action acquires a topological term which can account, at the quantum level, for dramatic differences between integer and half-integer spins. It seems however that no topological terms can arise [44] in the continuum limit of the two-dimensional, antiferromagnetic Heisenberg model. This is perhaps unfortunate, as the latter model is the strong-coupling limit of the two-dimensional Hubbard model at half-filling [16,17] , which is relevant for the description of the Cu-O planes in the newly discovered high-T_c superconducting materials [16,17], and topological terms in the continuum limit had been speculated by various authors to be at the origin of anyonic excitations in the Hubbard model as an explanation of high-T_c superconductivity [106]. However, the two-dimensional $SO(3)$ nonlinear σ-model is of interest in its own as a somewhat simplified version of the Skyrme model [14,15,91,92].

Actually, it was shown by Wilczek and Zee [108] that the action can be augmented by a new topological term, the so-called Hopf term which will be discussed shortly below. Such a term is harmless at the classical level, but, at the quantum level, it determines the spin and statistics of the solitonic solutions of the model (the "baby Skyrmions" of Wilczek and Zee). It is remarkable that it is precisely a nonlinear $(2+1)$ $SO(3)$ σ-model that seems to provide a good description of the low-energy effective action of He^3 films, and that a Hopf term seems to arise in a natural way in the derivation of the effective action [100]. As it happens ever so often in recent times, Condensed-Matter Physics offers concrete realizations of what are more or the less speculations and models in Quantum Field Theory.

As a final example in this Chapter, I will review how vortices and flux quantization in type-II superconductors can be derived by topological arguments. Before proceeding, however, I will discuss briefly some relevant topological invariants, namely the Pontrjagin index and the Hopf invariant, which are related to (and indeed classify) the homotopies $\pi_2(S^2)$ and $\pi_3(S^2)$. This will be done in the next Section.

2.2. THE PONTRJAGIN INDEX AND THE HOPF INVARIANT.

In this Section we will study two invariants related to the homotopy groups $\pi_2(S^2)$ and $\pi_3(S^2)$ respectively. We know from Ch. 1 that:

$$\pi_2(S^2) = \pi_3(S^2) = \mathbb{Z} \tag{2.4}$$

Maps in the corresponding homotopy classes are then characterized by *winding numbers*. The Pontrjagin index and the Hopf invariant allow us precisely to calculate the winding number of any such map.

Let us begin with $\pi_2(S^2)$. Denoting a map as in Eq. (2.1), we know from Ch. 1 that we may as well consider Φ as a map from the square $\mathbb{I} \times \mathbb{I}$ into S^2, provided the boundary $\partial(\mathbb{I} \times \mathbb{I})$ is all mapped into a single point of S^2. Also, $\vec{\phi}(x)$ in (2.1) may be thought of as a unit vector in $\mathbb{R}^3$, i.e.:

$$\vec{\phi}(x) \in \mathbb{R}^3, \ \ \vec{\phi}(x) \cdot \vec{\phi}(x) = 1 \ \forall x \in \mathbb{I} \times \mathbb{I} ; \tag{2.4}$$

Let then x^μ ($\mu = 1,2$; $x^1 = u$, $x^2 = v$ in the notation of Ch. 1) denote coordinates in $\mathbb{I} \times \mathbb{I}$. Then:

Theorem: The winding number associated with the map Φ is given by:

$$Q= \frac{1}{8\pi} \int\limits_0^1 dx^1 \int\limits_0^1 dx^2 \, P(x^1,x^2) \; ; \quad P(x^1,x^2)=: \epsilon^{\mu\nu} \, \vec{\phi} \cdot (\partial_\mu \vec{\phi} \times \partial_\nu \vec{\phi})$$

$$(2.5)$$

The quantity $P(x^1,x^2)/8\pi$ is called the **Pontrjagin density** associated with the map Φ, and Q is called the **Pontrjagin index.**

In order to prove the theorem, let's introduce, on $\mathbb{I} \times \mathbb{I}$, the two-form:

$$\mathbb{A}=[\vec{\phi} \cdot \frac{\partial \vec{\phi}}{\partial x^1} \times \frac{\partial \vec{\phi}}{\partial x^2}] \, dx^1 \wedge \, dx^2 \equiv \tfrac{1}{2} \epsilon_{ijk} \, \phi^i \partial_\mu \phi^j \partial_\nu \phi^k \, dx^\mu \wedge \, dx^\nu$$

$$(2.6)$$

(i,j,k= 1,2,3). Then:

$$Q= \frac{1}{4\pi} \int\limits_{\mathbb{I} \times \mathbb{I}} \mathbb{A} \qquad (2.7)$$

Now, denoting as usual [1,30] by a superscript "*" the pull-back of forms:

$$\mathbb{A}= \Phi^*(\tfrac{1}{2}\epsilon_{ijk}\phi^i d\phi^j \wedge d\phi^k)= \Phi^*\eta \qquad (2.8)$$

where: $\eta = \tfrac{1}{2}\epsilon_{ijk}\phi^i d\phi^j \wedge d\phi^k$ is a two-form on $\mathbf{S}^2$. Then:

$$Q= \frac{1}{4\pi} \int\limits_{\mathbb{I}\times\mathbb{I}} \mathbb{A}= \frac{1}{4\pi} \int\limits_{\mathbb{I}\times\mathbb{I}} \Phi^*\eta = \frac{1}{4\pi} \int\limits_{\Phi(\mathbb{I}\times\mathbb{I})} \eta \qquad (2.9)$$

Taking into account the constraint (2.4), η turns out to be the solid angle on the two-sphere. The last expression in (2.9) is then nothing but the *degree* [26,30,43] of the map Φ, i.e. the number of times the map winds around the target space S^2, and this proves the theorem. □.

As already stated in Sect. 2.1, any quantity that depends only on the winding number of a map is a homotopy invariant, i.e. it does not change under smooth deformations of the map itself. That this is so for the Pontrjagin index can be seen directly in the following manner. By varying $\vec{\phi}(x)$ infinitesimally:

$$\vec{\phi}\to \vec{\phi}+ \delta\vec{\phi} \qquad (2.10)$$

we find, with some long but simple algebra, and neglecting boundary terms (i.e. we assume: $\delta\vec{\phi}=0$ on $\partial(\mathbb{I}\times\mathbb{I})$)):

$$\delta Q= \frac{3}{8\pi} \int d^2x \; \epsilon^{\mu\nu}\delta\vec{\phi}\cdot(\partial_\mu\vec{\phi}\times\partial_\nu\vec{\phi}) \qquad (2.11)$$

But, by virtue of the constraint (2.4) on $\vec{\phi}$:

$$\vec{\phi} \cdot \delta\vec{\phi} = \vec{\phi} \cdot \partial_\mu \vec{\phi} = 0 \; \forall \mu \qquad (2.12)$$

Therefore, $\delta\vec{\phi}$ and the partial derivatives $\partial_\mu \vec{\phi}$ are all orthogonal to $\vec{\phi}$. Hence, they are coplanar, and the triple product in (2.12) vanishes. So: $\delta Q=0$, and this is a direct proof that Q is indeed a homotopy invariant.□

Let us turn now to $\pi_3(S^2)$, and consider a map: $\Phi: S^3 \rightarrow S^2$. Let Ω be a normalized volume-form on S^2, i.e. a (closed but not exact) two-form such that:

$$\int_{S^2} \Omega = 1 \qquad (2.13)$$

For example, we may take: $\Omega = \eta/4\pi$, with η defined as in (2.8) (and, again, $\vec{\phi} \in S^2$). Explicitly, in that case:

$$\Omega = \frac{1}{4\pi} (\phi^1 d\phi^2 \wedge d\phi^3 + \phi^3 d\phi^1 \wedge d\phi^2 + \phi^2 d\phi^3 \wedge d\phi^1) \qquad (2.14)$$

or, in spherical polar coordinates ($\phi^1 = \sin\theta\cos\phi$, $\phi^2 = \sin\theta\sin\phi$, $\phi^3 = \cos\theta$):

$$\Omega = \frac{1}{4\pi} \sin\theta d\theta \wedge d\phi \qquad (2.15)$$

Consider now the pull-back $\Phi^*\Omega$ of Ω to S^3 via the map Φ. As the second cohomology group of S^3 vanishes [30], all closed two-forms are exact, and there exists a one-form ω such that: $\Phi^*\Omega = d\omega$. Then:

Definition: The *Hopf invariant* associated with the map Φ is given by:

$$H(\Phi) = \int_{S^3} \omega \wedge d\omega \qquad (2.16)$$

It is easy to show that the Hopf invariant is independent of the choice of ω. Indeed, if ω' is another one-form such that: $\Phi^*\alpha = d\omega'$, then $\omega - \omega'$ is a closed (actually exact) form, and:

$$\int_{S^3} (\omega - \omega') \wedge d\omega \equiv \int_{S^3} d\left((\omega' - \omega) \wedge \omega\right) = 0 \qquad (2.17)$$

by Stokes' theorem ☐.

The main property of the Hopf invariant is however contained in the following:

Theorem: The value of the Hopf invariant depends only on the homotopy class of the map Φ, i.e. $H(\Phi)$ is a **homotopy invariant.**

We will not prove the theorem here, but refer to the literature [26,43,54] for the proof, which, although not difficult, is rather lengthy.

It turns out also that, with the chosen normalization of the volume-form Ω, the Hopf invariant is always an integer. Therefore, it can be used to calculate the winding number associated with the map Φ.

Exercise. The Hopf invariant of the Hopf fibration.

The Hopf fibration: $S^1 \Rightarrow S^3 \Rightarrow S^2$ has been discussed in Ch.1, § 1.6. We want to evaluate now the Hopf invariant associated with the projection $\pi : S^3 \rightarrow S^2$ defined in Eq.(1.75).

It will be convenient to work on coordinate patches. S^2 can be covered as:

$$S^2 = V_0 \cup V_1 \qquad (2.19\text{-a})$$

where:

$$V_0 = S^2 - (0,0,-1); \quad V_1 = S^2 - (0,0,1) \qquad (2.19\text{-b})$$

On V_0 we can use the stereographic chart (corresponding to the stereographic projection on the equatorial plane from the South pole $(0,0,-1)$):

$$x = \frac{\phi^1}{1 + \phi^3}, \quad y = \frac{\phi^2}{1 + \phi^3} \qquad (2.20)$$

which inverts to:

$$\phi^1 = \frac{2x}{1 + r^2}, \quad \phi^2 = \frac{2y}{1 + r^2}, \quad \phi^3 = \frac{1 - r^2}{1 + r^2}; \quad r^2 = x^2 + y^2 \qquad (2.21)$$

A similar stereographic chart can be constructed for V_1, but we will not need it here. Direct calculation shows that the volume-form (2.14) becomes, in the chart (2.20):

$$\Omega = \frac{1}{\pi} \frac{dx \wedge dy}{(1+r^2)^2} \tag{2.22}$$

S^3, with complex coordinates α, β, $|\alpha|^2 + |\beta|^2 = 1$ (see Sect. 1.6), can be covered in a similar way by two open sets U_0 and U_1:

$$U_0 = S^3 - (\alpha = 0) , \quad U_1 = S^3 - (\beta = 0) \tag{2.23}$$

One can prove easily that U_0 (U_1) projects down to V_0 (V_1) under the projection (1.75). In terms of the stereographic coordinates for V_0, U_0 can be given coordinates (x, y, χ) with: $0 \leq \chi < 2\pi$, and:

$$\alpha = \frac{e^{i\chi}}{\sqrt{1+r^2}} ; \quad \beta = \frac{ze^{i\chi}}{\sqrt{1+r^2}} ; \quad z = x + iy \tag{2.24}$$

Therefore, on $U_0 = \pi^{-1}(V_0)$:

$$\pi^*\Omega = \frac{1}{\pi} \frac{dx \wedge dy}{(1+r^2)^2} = d\omega \tag{2.25}$$

where the one-form ω can be chosen [14] as:

$$\omega = \frac{1}{2\pi} \left(\frac{xdy - ydx}{1+r^2} + d\chi \right) \tag{2.26}$$

(note that, χ being an angle, $d\chi$ is a closed but not exact one-form). The one-form (2.26) can be shown [14] to be the expression in local coordinates of the globally defined one-form:

$$\omega = \frac{i}{4\pi} \, \text{Tr} \, (\sigma_3 s^{-1} ds) \qquad (2.27)$$

with s given by Eq. (1.60).

Direct computation shows further that, in local coordinates:

$$\omega \wedge d\omega = \frac{1}{2\pi^2} \frac{dx \wedge dy \wedge d\chi}{(1+r^2)^2} \qquad (2.28)$$

It is now easy to evaluate the integral (2.16), and we find:

$$H(\pi) = \frac{1}{2\pi^2} \int_{-\infty}^{+\infty} dx \int_{-\infty}^{+\infty} dy \int_0^{2\pi} d\chi \, \frac{1}{(1+r^2)^2} = 1 \qquad (2.29)$$

proving that the Hopf invariant associated with the Hopf fibration is equal to one.

We want now to discuss briefly another, more geometrical, aspect of the Hopf invariant.

With reference, e.g., to the coordinatization (2.24), it is quite clear that the inverse image under the Hopf map π of any point in S^2 will be a closed curve (actually a circle) in S^3. Just as in the case of S^2, we can construct stereographic projections from any point of S^3 onto $\mathbb{R}^3$, with the projection point being identified with the "point at infinity" of $\mathbb{R}^3$ (i.e. projections from S^3 to the one-point compactification of $\mathbb{R}^3$). By setting: $\alpha=x^1+ix^2$, $\beta=x^3+ix^4$ ($\sum(x^i)^2=1$) and denoting by ζ^i, i=1,2,3, the Cartesian coordinates in $\mathbb{R}^3$, one such projection from, say, the point $(0,0,0,-1)$ in S^3 will be:

$$\zeta^i = \frac{x^i}{1+x^4} , \ i=1,2,3; \quad x^4 \neq -1 \tag{2.30}$$

which inverts in a form similar to (2.21).Closed curves in S^3 can be then visualized more easily as closed curves in $\mathbb{R}^3$ via the stereographic projection (2.30).

Given any two closed curves A and B in $\mathbb{R}^3$ (or, via the inverse of (2.30), in S^3), their *linking number* can be defined as follows [26,43,54]: take a smooth surface D in $\mathbb{R}^3$ with boundary A such that B intersects D transversally. The orientation of D (i.e. that of the positive normal) will be fixed by the parametrization of A with the usual conventions, i.e. in such a way that A is traced counterclockwise with respect to the direction of the positive normal. The linking number is then defined as:

$$\text{Link}(A,B)= \sum_{D\cap B} (\pm 1) \tag{2.31}$$

where the sum is over the points of intersection of B with D and the sign is taken according to whether B intersects D in the same or in the opposite direction of the positive normal to D. That Link(A,B) = Link(B,A) and that the result is independent of the choice of the surface D requires a rather lengthy proof which will be omitted here [26].

Now, it turns out (again, we will omit the proof, and refer to the literature [26,43,54] for it) that, in general, given a map $\Phi: S^3 \to S^2$, *the Hopf invariant of the map is equal to the linking number of the inverse images of any two distinct points of* S^2. Through this theorem the Hopf invariant acquires therefore a neat geometrical interpretation.

Note. The proof of the theorem requires actually a further technical specification, namely that the chosen points in S^2 be **regular** [26] values of the map. We will not insist however here on these technical details.

Let us close this Section by giving an explicit example of calculation of a linking number, referring again to the Hopf fibration and to the projection map π. Consider the two points : p=(1,00) and q= (-1,0,0) in S^2. Their inverse images in S^3 are the circles (cfr. (2.24)):

$$C_p: \alpha=\beta= \frac{e^{i\chi}}{\sqrt{2}} \; ; \; C_q: \alpha=-\beta= \frac{e^{i\chi}}{\sqrt{2}} \qquad (2.32)$$

By stereographic projection we obtain, in $\mathbb{R}^3$, the curves:

$$E_p: \zeta^1 = \zeta^3 = \frac{\cos\chi}{\sqrt{2}+\sin\chi} \ ; \ \zeta^2 = \frac{\sin\chi}{\sqrt{2}+\sin\chi} \qquad (2.33\text{-a})$$

and:

$$E_q: \zeta^1 = -\zeta^3 = \frac{\cos\chi}{\sqrt{2}-\sin\chi} \ ; \ \zeta^2 = \frac{\sin\chi}{\sqrt{2}-\sin\chi} \qquad (2.33\text{-b})$$

E_p and E_q are two ellipses lying on the mutually orthogonal planes $\zeta^1 = \pm\zeta^3$. An elementary calculation shows that they link exactly once, in agreement with the already known fact that $H(\pi)=1$.

2.3. THE SO(3) NONLINEAR SIGMA-MODELS IN ONE AND TWO SPACE DIMENSIONS.

We shall consider in this Section the $SO(3)$ nonlinear σ-models [84,115] in one and two space dimensions. With reference to Sect. 2.1, we shall assume here $M = \mathbb{R}^{d+1}$, d=1, or 2, and $T = S^2$. The field space will be therefore the homogeneous space $SO(3)/SO(2)$ (see Ch.1),which can be identified also with the complex projective plane $\mathbb{CP}^1$.

The field $\vec{\phi}$ will be a unit vector in $\mathbb{R}^3$, i.e.:

$$\vec{\phi}(x) \in \mathbb{R}^3 \; \forall x; \; \vec{\phi}(x) \cdot \vec{\phi}(x) \equiv 1 \qquad (2.34)$$

According to what has just been said, we shall consider either a (1+1) (d=1) or a (2+1) (d=2) dimensional flat Minkowskian spacetime, with metric: $g_{\mu\nu} = \text{diag}(1,-1)$ (d=1) or: $g_{\mu\nu} = \text{diag}(1,-1,-1)$ (d=2). We will assume the system to be described by the Lagrangian density:

$$\mathcal{L} = \frac{1}{2g} (\partial_\mu \vec{\phi}) \cdot (\partial^\mu \vec{\phi}) \qquad (2.35)$$

where g is a coupling constant.

The Lagrangian (2.35), as such, describes a free field theory. However, the constraint (2.34) turns the theory into a nonlinear one. Indeed, the constraint has to be incorporated in the action functional via a Lagrange multiplier, and we find for the action:

$$S[\vec{\phi}]=\frac{1}{g}\int d^{d+1}x \left\{\tfrac{1}{2}(\partial_\mu\vec{\phi}\cdot\partial^\mu\vec{\phi})+ \lambda(x)(\vec{\phi}(x)\cdot\vec{\phi}(x)-1)\right\} \qquad (2.36)$$

where $\lambda(x)$ is the Lagrangian multiplier.

The field equations can be derived in a straightforward way from (2.36), and they read:

$$[\partial_\mu\partial^\mu+ \lambda]\vec{\phi}(x)= 0 \qquad (2.37)$$

We can now solve for the constraint. Indeed, from (2.37):

$$\lambda(x) \equiv \lambda(x)(\vec{\phi}(x)\cdot\vec{\phi}(x))= -\vec{\phi}(x)\cdot(\partial^\mu\partial_\mu\vec{\phi}) \qquad (2.38)$$

Therefore, the field equations become:

$$\partial^\mu\partial_\mu\vec{\phi}- (\vec{\phi}\cdot\partial^\mu\partial_\mu\vec{\phi})\vec{\phi}= 0 \qquad (2.39)$$

Note that every solution of (2.39) satisfies automatically the constraint (2.34). (2.39) is a **nonlinear** set of field equations, and it may admit of *solitonic* (i.e. finite-energy, localized) solutions. The latter were first studied in d=2 by Belavin and Polyakov (see Ref. [84] for a review).

The transition to the Hamiltonian formalism can be done by

standard techniques [1,84]. The conjugate momenta are defined as:

$$\vec{\pi}(x)= \frac{\partial \mathcal{L}}{\partial(\partial_0\vec{\phi})} = \frac{1}{g}\partial_0\vec{\phi} \tag{2.40}$$

The Hamiltonian density is therefore:

$$\mathcal{H}(x)= \vec{\pi}\cdot\partial_0\vec{\phi}- \mathcal{L}= \frac{1}{2}\,[g\vec{\pi}^2(x)+\frac{1}{g}\partial_a\vec{\phi}(x)\cdot\partial_a\vec{\phi}(x)] \tag{2.41}$$

($\partial_a \equiv \partial/\partial x^a$, where a is a space index (a=1 (d=1) or a=1,2 (d=2)).

The total energy of a field configuration will be :

$$\mathcal{E}= \int \mathcal{H}(x)\; d^d x \tag{2.42}$$

In particular, for static solutions ($\vec{\pi}(x) \equiv 0$), the total energy will be:

$$\mathcal{E}= \frac{1}{2g}\int d^d x \; (\partial_a\vec{\phi}\cdot\partial_a\vec{\phi}) \tag{2.43}$$

Quite clearly, $\mathcal{E} \geq 0$, and $\mathcal{E}=0$ (the "classical vacua") correspond to:

$$\vec{\phi} \equiv \; const. \tag{2.44}$$

Remark: The theory is fully $SO(3)$ (actually $O(3)$) invariant under the action of $SO(3)$ on $\vec{\phi}$. Any solution of the form (2.44) will break the symmetry down to $SO(2)$, the isotropy subgroup of the constant value of $\vec{\phi}$.

Let us concentrate now on the one-dimensional model. Hence, we will assume d=1 for the time being. Consider first the case of static solution of the field equations. Finiteness of the total energy (2.43) requires $\vec{\phi}$ to tend to a constant value at space infinity. Space is therefore compactified (See Sect. 2.1) to a circle S^1, and the fields are maps from S^1 to S^2. But we know that $\pi_1(S^2)=0$. Therefore, there are no topologically nontrivial static solutions of the model.

The situation is different when we consider genuinely time-dependent solutions. If we require again the fields to go to a constant value at **spacetime** infinity, the fields will be now maps from S^2 to S^2. Therefore, the space of fields will break into homotopy sectors labeled by $\pi_2(S^2)=\mathbb{Z}$, and indexed by the Pontrjagin index which we studied in the previous Section. This opens up the interesting possibility of "augmenting" the action of the model by a term proportional to the homotopy invariant (the Pontrjagin index) which classifies the different sectors, or, equivalently, of adding to the Lagrangian a term proportional to the Pontrjagin density.

Let's consider then the modified Lagrangian:

$$\mathcal{L} = \frac{1}{2g}\,(\partial_\mu\vec{\phi})\cdot(\partial^\mu\vec{\phi}) - \frac{\theta}{8\pi}\,\epsilon^{\mu\nu}\vec{\phi}\cdot(\partial_\mu\vec{\phi}\times\partial_\nu\vec{\phi}) \qquad (2.45)$$

where θ is an arbitrary real parameter.

The second term in (2.45) is also known in the literature as the **Wess-Zumino** [103,116] term. Written explicitly, the contribution of the Wess-Zumino term to the action is:

$$S_{WZ} = -\frac{\theta}{4\pi} \int d^2x \, \vec{\phi} \cdot (\partial_0\vec{\phi} \times \partial_1\vec{\phi}) \qquad (2.46)$$

Let us reexamine now the transition to the Hamiltonian formalism in the presence of a Wess-Zumino term. The conjugate (density of) momenta associated to the field $\vec{\phi}$ are now given by:

$$\vec{p}(x) = \frac{\partial \mathcal{L}}{\partial(\partial_0\vec{\phi}(x))} = \frac{1}{g} \partial_0\vec{\phi} + \frac{\theta}{4\pi} \vec{\phi} \times \partial_1\vec{\phi} \qquad (2.47)$$

The standard Poisson brackets will be:

$$[\![\phi^i(x), p^j(y)]\!] = \delta^{ij}\delta(x-y)$$

$$\qquad (2.48)$$

$$[\![\phi^i(x),\phi^j(y)]\!] = [\![p^i(x),p^j(y)]\!] = 0$$

(i,j=1,2,3). The Hamiltonian density associated with (2.45) will be:

$$\mathcal{H} = \vec{p} \cdot \partial_0\vec{\phi} - \mathcal{L} = \frac{1}{2} g \left(\vec{p} - \frac{\theta}{4\pi} \vec{\phi} \times \partial_1\vec{\phi}\right)^2 + \frac{1}{2g} |\partial_1\vec{\phi}|^2 \qquad (2.49)$$

Rotations in the field space are canonically generated by:

$$\vec{l}(x) = \vec{\phi}(x) \times \vec{p}(x) \qquad (2.50)$$

and, indeed:

$$[l^i(x), \phi^j(y)] = \epsilon^{ijk}\phi^k(x)\delta(x-y) \qquad (2.51)$$

(with similar Poisson bracket relations holding between the components of $\vec{l}$ and $\vec{p}$). Now, the constraint on $\vec{\phi}$ implies:

$$0 = \vec{\phi} \cdot \partial_\mu \vec{\phi} \Rightarrow \vec{\phi} \cdot \vec{p} = 0 \qquad (2.52)$$

Also, by construction: $\vec{\phi} \cdot \vec{l} = 0$. All in all, this implies:

$$\vec{p} = -\vec{\phi} \times \vec{l} \qquad (2.53)$$

Therefore, the Hamiltonian (2.49) can be rewritten as:

$$\mathcal{H} = \frac{1}{2} g \, |\vec{l} + \frac{\theta}{4\pi} \partial_1 \vec{\phi}|^2 + \frac{1}{2g} |\partial_1 \vec{\phi}|^2 \qquad (2.54)$$

With the proper identification of the parameters, namely:

$$\frac{1}{g^2} = \frac{S}{2} \;;\; \theta = 2\pi S \tag{2.55}$$

this is precisely the Hamiltonian that describes [2] the long wavelength limit of the antiferromagnetic Heisenberg chain with spin S. With reference to what we know already about the Pontrjagin index, we will have, for any field configuration that tends to some constant value at spacetime infinity:

$$S_{WZ} = -2\pi S \times \text{(integer)} \tag{2.56}$$

At the quantum level (see the discussion of Sect. 2.1) this will imply definite differences between integer and half-odd-integer spins. We will not insist on this point here, as related problems will be discussed in more detail in Ch. 3.

Let us turn now to the two-dimensional (d=2) case, and look again for static solutions of the field equations with $0 < \mathcal{E} < \infty$. The finite-energy condition implies:

$$\lim_{r \to \infty} \vec{\phi}(\vec{x}) = \text{const.} = \vec{\phi}_0; \;\; \lim_{r \to \infty} r \, |\, \nabla \vec{\phi}\, | = 0 \tag{2.57}$$

$(r = |\,\vec{x}\,|)$, i.e. that $\vec{\phi}$ tends to a constant value at infinity rapidly enough to make the integral (2.43) to converge. Also, we will set g=1 in what follows.

The fact that $\vec{\phi}$=const. at infinity allows us to perform a one-point compactification of $\mathbb{R}^2$ to S^2. It follows then that *physical (i.e. finite-energy) static field configurations correpond now to maps of S^2 onto S^2*, which are classified by $\pi_2(S^2)$ and by the Pontrjagin index:

$$n \equiv Q = \frac{1}{8\pi} \int d^3x \; \epsilon^{ij} \vec{\phi} \cdot (\partial_i \vec{\phi} \times \partial_j \vec{\phi})$$ (2.58)

(i,j=space indices).

Solitons in the nonlinear $SO(3)$ model have been studied by Belavin and Polyakov. They have shown (see Ref.[84] for details) that the identity:

$$\int d^3x \; | \partial_\mu \vec{\phi} \pm \epsilon_{\mu\nu} \vec{\phi} \times \partial_\nu \vec{\phi} |^2 \geq 0 \quad \forall \vec{\phi}$$ (2.59)

leads to the inequality:

$$\mathcal{E} \geq 4\pi \, | Q |$$ (2.60)

Solutions minimizing the energy will correspond then to $\mathcal{E}=4\pi \, | Q |$, and will solve the *first-order* equation(s):

$$\partial_\mu \vec{\phi} = \pm \epsilon_{\mu\nu} \vec{\phi} \times \partial_\nu \vec{\phi} \qquad (2.61)$$

with the lower sign corresponding to $Q > 0$, the upper one to $Q < 0$ (See again ref. [84] for details). Note that, as under $\vec{\phi} \to -\vec{\phi}$ the l.h.s. of (2.61) is **odd**, while the r.h.s. is **even**, solutions with a given winding number will produce in an obvious way also solutions for the **opposite** winding.

Exercise . Show that a solution of (2.61) corresponding to Q=1 is given by:

$$\vec{\phi}(\vec{x}) = (\hat{x} \sin f, \cos f) \qquad (2.62)$$

where

$$\hat{x} = \frac{\vec{x}}{r} ; \quad r = |\vec{x}| ; \quad f = f(r) \text{ with: } \sin f = \frac{2r}{1+r^2}, \cos f = \frac{r^2-1}{r^2+1} \qquad (2.63)$$

Note that, as r grows from 0 to infinity, f(r) decreases monotonically from $f(0)=\pi$ to $f(\infty)=0$. This shows intuitively that (2.62) should cover S^2 exactly once.

Hint: Show first that Eq. (2.61) (with the minus sign) leads to:

$$\epsilon^{ij} \vec{\phi} \cdot \partial_i \vec{\phi} \times \partial_j \vec{\phi} \equiv \partial_i \vec{\phi} \cdot \partial_i \vec{\phi} \equiv \sum_i |\nabla \phi_i|^2 \qquad (2.64)$$

Hence:

$$Q = \frac{1}{8\pi} \int d^2x \sum_i \mid \nabla \phi_i \mid^2 \qquad (2.65)$$

On the other hand:

$$\frac{df}{dr} = -\frac{2}{r^2+1} \qquad (2.66)$$

and hence:

$$\sum_i \mid \nabla \phi_i \mid^2 = \frac{8}{(r^2+1)^2} \qquad (2.67)$$

Substituting (2.67) back into (2.65) we obtain Q=1. $\square$.

2.4. THE d=2 SO(3) NONLINEAR σ-MODEL AS A GAUGE THEORY.

There is an interesting representation [15,17,84,108,115] of the SO(3) nonlinear σ-model which allows it to be interpreted, to some extent, as a U(1) gauge theory.

We will employ here a representation of the Hopf map $S^3 \to S^2$ which is somewhat different from that employed in Ch. 1, but more useful for computational purposes. To this effect, let's consider a normalized spinor:

$$z = \begin{vmatrix} z_1 \\ z_2 \end{vmatrix}; \ z_i = z_i(x) \in \mathbb{C}; \ z^\dagger z = 1 \Leftrightarrow |z_1|^2 + |z_2|^2 = 1 \quad (2.68)$$

With the normalization fixed, z spans the three-sphere S^3, which is the group manifold of SU(2). An alternative way of exhibiting the Hopf map is obtained by defining:

$$\vec{\phi} =: z^\dagger \vec{\sigma} \ z \equiv (z_\alpha{}^* z_\beta) \vec{\sigma}_{\alpha\beta} \quad (2.69)$$

where $\vec{\sigma} \equiv (\sigma_1, \sigma_2, \sigma_3)$ are the Pauli matrices. Using the identity:

$$\vec{\sigma}_{\alpha\beta} \cdot \vec{\sigma}_{\gamma\delta} \equiv 2\delta_{\alpha\delta}\delta_{\beta\gamma} - \delta_{\alpha\beta}\delta_{\gamma\delta} \quad (2.70)$$

one proves at once that:

$$\vec{\phi}\cdot\vec{\phi} \equiv z^{\dagger}z \qquad (2.71)$$

and hence that $\vec{\phi} \in S^2 \Leftrightarrow z \in S^3$. Also, as, from (2.69):

$$\phi^1 + i\phi^2 = 2z^{1*}x^2 \; ; \; \phi^3 = |z^1|^2 - |z^2|^2 \qquad (2.72)$$

one obtains at once, with reference to the representation (1.60) of the elements of $SU(2)$, the identification: $z^1 = \alpha$, $z^2 = \beta$. The $U(1)$ subgroup leaving ϕ unchanged is now represented by the (local) $U(1)$ transformation:

$$z(x) \rightarrow \exp[i\theta(x)]z(x) \qquad (2.73)$$

In this way, we have managed to go from an S^2-valued field $(\vec{\phi})$ to an $SU(2)$ valued one (z) which appears to have the $U(1)$ gauge freedom (2.73).

Consider next the Lagrangian of Eq. (2.35) (with g=1). With some algebra, we find:

$$\frac{1}{2}\partial_\mu\vec{\phi}\cdot\partial^\mu\vec{\phi} \equiv [\partial_\mu(z_\alpha{}^*z_\beta)][\partial^\mu(z_\beta{}^*z_\alpha)] \qquad (2.74)$$

where the constraint $z^{\dagger}z=1$ has been used in order to obtain the result on the r.h.s. Again from the constraint we obtain:

$$1 = z^{\dagger}z \equiv z_\alpha{}^*z_\alpha \Rightarrow z_\alpha{}^*\partial_\mu z_\alpha + z_\alpha\partial_\mu z_\alpha{}^* = 0 \; \forall \, \mu \qquad (2.75)$$

i.e.that $z^\dagger \partial_\mu z$ is a **purely imaginary** quantity.

With this in mind we obtain, again with some algebra:

$$[\partial_\mu(z_\alpha{}^* z_\beta)][\partial^\mu(z_\beta{}^* z_\alpha)] \equiv 2[(\partial z_\alpha)(\partial^\mu z_\alpha{}^*) + (z_\alpha{}^* \partial_\mu z_\alpha)(z_\beta{}^* \partial^\mu z_\beta)] \equiv$$

$$\equiv 2 \, (\partial_\mu z - (z^\dagger \partial_\mu z)z)^\dagger (\partial^\mu z - (z^\dagger \partial^\mu z)z) \qquad (2.76)$$

Altogether:

$$\tfrac{1}{2} \, \partial_\mu \vec{\phi} \cdot \partial^\mu \vec{\phi} = 2(\partial_\mu z - (z^\dagger \partial_\mu z)z)^\dagger (\partial^\mu z - (z^\dagger \partial^\mu z)z) \qquad (2.77)$$

Let's introduce now the $\mathbb{U}(1)$ gauge potential:

$$A_\mu =: -iz^\dagger \partial_\mu z \qquad (2.78)$$

In view of (2.76) A_μ is a **real** quantity, and, under the gauge transformation (2.73):

$$A_\mu \rightarrow A_\mu + \partial_\mu \theta \qquad (2.79)$$

The associated field strength $F_{\mu\nu} = \partial_\mu A_\nu - \partial_\nu A_\mu$ is of course gauge invariant. As such, it should be projectable to a well defined tensor field on $\mathbb{S}^2$ while, A_μ is not, and lives genuinely on $\mathbb{S}^3$.

We can rewrite then:

$$\partial_\mu z - (z^\dagger \partial_\mu z)z \equiv D_\mu z \; ; \quad D_\mu =: \partial_\mu - iA_\mu \qquad (2.80)$$

D_μ is therefore the **covariant derivative** [30,31,40,77,78] associated with the gauge field A_μ, and we have, as usual:

$$z \to z \cdot \exp[i\theta] \Rightarrow (D_\mu z) \to \exp[i\theta](D_\mu z) \qquad (2.81)$$

The field Lagrangian can be written now as:

$$\mathcal{L} = \tfrac{1}{2} \partial_\mu \vec{\phi} \cdot \partial^\mu \vec{\phi} \equiv 2(D_\mu z)^\dagger (D^\mu z) \qquad (2.82)$$

We have thus succeeded in mapping the $SO(3)$ nonlinear σ-model into a $U(1)$ gauge theory, whose field Lagrangian is given by the r.h.s. of (2.82), and with the additional, gauge-invariant constraint:

$$z^\dagger(\vec{x},t) \cdot z(\vec{x},t) \equiv 1 \qquad (2.83)$$

Let us stress however that *the gauge potential A_μ has no independent dynamics.* Indeed, varying the Lagrangian on the r.h.s. of (2.82) w.r.t. A_μ yields simply the definition of A_μ in therms of the field z. The present theory therefore is not a genuine gauge theory, although it has been possible to "manufacture" a gauge potential out of it.

With the gauge potential (2.79) we can associate the one-form:

$$\mathbb{A}= A_{\mu}dx^{\mu}= -iz^{\dagger}dz \qquad (2.84)$$

Comparing with the discussion of Sect. 2.2, we see easily that $\mathbb{A}$ is related to the one-form ω defined in Eq. (2.27) by:

$$\mathbb{A}= -2\pi\omega \qquad (2.85)$$

Also, the field strength associated with (2.78) will determine the two-form:

$$\mathbb{F}= d\mathbb{A}=\frac{1}{2} F_{\mu\nu}dx^{\mu}\wedge dx^{\nu}; \quad F_{\mu\nu}= \partial_{\mu}A_{\nu} - \partial_{\nu}A_{\mu} \qquad (2.86)$$

and, of course: $\mathbb{F}= -2\pi d\omega$.

Let's study now in more detail the field strength $F_{\mu\nu}$. It follows from its definition that:

$$F_{\mu\nu}= \partial_{\mu}A_{\nu}- \partial_{\nu}A_{\mu} \equiv -i(\partial_{\mu}z^{\dagger}\cdot\partial_{\nu}z- \partial_{\nu}z^{\dagger}\cdot\partial_{\mu}z) \qquad (2.87)$$

It is left as an exercise to show explicitly that the r.h.s. of (2.87) is actually gauge-invariant.

A long but straightforward algebra shows that, with the definition.(2.69) of $\vec{\phi}$ in terms of z:

$$F_{\mu\nu} \equiv -\frac{1}{2}\vec{\phi}\cdot(\partial_{\mu}\vec{\phi}\times\partial_{\nu}\vec{\phi}) \qquad (2.88)$$

as it was to be expected. This exhibits the gauge-invariance of the field strength in an explicit way.

The **dual** field strength, which is a Lorentz **vector** in (2+1) spacetime dimensions, is given by:

$$j^{\mu} =: \frac{1}{2} \epsilon^{\mu\nu\lambda} F_{\nu\lambda} \equiv \frac{1}{2} \epsilon^{\mu\nu\lambda} \vec{\phi} \cdot (\partial_{\nu}\vec{\phi} \times \partial_{\lambda}\vec{\phi}) \qquad (2.89)$$

Had we defined j^{μ} directly using the r.h.s. of (2.89), it would have appeared as a current associated with the original model. It is actually a **conserved** current, as:

$$\partial_{\mu}j^{\mu} = \frac{1}{2} \epsilon^{\mu\nu\lambda} \partial_{\mu}\vec{\phi} \cdot (\partial_{\nu}\vec{\phi} \times \partial_{\lambda}\vec{\phi}) \qquad (2.90)$$

and the r.h.s. vanishes identically, the three vectors being coplanar because of the constraint on $\vec{\phi}$. So: $\partial_{\mu}j^{\mu}=0$ irrespective of the field dynamics, and j^{μ} appears as a **topological current** associated with the model. The corresponding conserved charge:

$$\tilde{Q} = \int d^3x \, j^0(x) \qquad (2.91)$$

satisfies: $\tilde{Q} = 4\pi Q$, where Q is the Pontrjagin index . *The Pontrjagin invariant can be viewed then as the topological charge associated with the conserved topological current j^{μ}.*

A general field configuration z(x) will be, via the projection map (2.69), a map from S^3 to S^2. We can then construct the Hopf invariant associated with the field z(x). In terms of the gauge potential A we obtain (cfr. Sect. 2.2):

$$H(\vec{\phi})= \frac{1}{4\pi^2} \int_{S^3} A \wedge dA \qquad (2.92)$$

Explicitly:

$$H(\vec{\phi})= -\frac{1}{4\pi^2} \int_{S^3} \epsilon^{\mu\nu\lambda}(z^\dagger \cdot \partial_\mu z)[(\partial_\nu z^\dagger) \cdot (\partial_\lambda z)] \qquad (2.93)$$

Remark. In view of the antisymmetry of the Ricci tensor, all terms in the integrand of (2.93) will contain a time derivative. Therefore the Hopf invariant (2.93) will vanish for *static* solutions and, in general, for static field configurations.

As already stated in Sect. 2.1, topological invariants can be expressed as integrals over the (compactified) spacetime manifold of (local) total divergences. Indeed, this is almost obvious from the differential-form expression of both the Pontrjagin index and the Hopf invariant. Both are expressed as integrals over appropriate manifolds (S^2 and S^3 respectively) of forms of maximal rank. As such, the forms are closed but not exact (as [30] $H^2(S^2)= H^3(S^3)=\mathbb{R}$). Now, being closed and hence, by Poincare' Lemma [30] locally exact is simply another way of saying that, in local coordinates and with the Lorentz metric, the coefficient of a form of maximal rank can be written as a total divergence. However, for the sake of exercise, we will derive this result explicitly for the Hopf invariant.

With reference to (2.93), let us set (cfr. Sect. 2.2):

$$z_1 = x^1 + ix^2; \; z_2 = x^3 + ix^4; \; \sum_{i=1}^{4} (x^i)^2 = 1 \qquad (2.94)$$

Let us work on the open sets $x^4 \neq 0$ on $\mathbf{S}^3$. Then the constraint on the x^i's yields:

$$dx^4 = -\frac{1}{x^4} (x^1 dx^1 + x^2 dx^2 + x^3 dx^3) \qquad (2.95)$$

and a long but otherwise straightforward algebra leads to:

$$\epsilon^{\mu\nu\lambda} A_\mu \partial_\nu A_\lambda = -\frac{2}{x^4} \epsilon^{\mu\nu\lambda} \partial_\mu x^1 \partial_\nu x^2 \partial_\lambda x^3 \qquad (2.96)$$

The stereographic chart (2.30) inverts to:

$$x^i = \frac{2\zeta^i}{1 + \vec{\zeta}^2} , \; i=1,2,3; \; x^4 = \frac{1 - \vec{\zeta}^2}{1 + \vec{\zeta}^2} ; \; \vec{\zeta}^2 = \sum_{i=1}^{3} (\zeta^i)^2 \qquad (2.97)$$

and, in terms of the stereographic coordinates (2.97), (2.96) can be rewritten as:

$$\epsilon^{\mu\nu\lambda}A_\mu\partial_\nu A_\lambda = -\frac{16}{(1+\vec{\zeta}^2)^2}\epsilon^{\mu\nu\lambda}\partial_\mu\zeta^1\partial_\nu\zeta^2\partial_\lambda\zeta^3 \qquad (2.98)$$

Finally, (2.98) can be rewritten explicitly as a total divergence:

$$\epsilon^{\mu\nu\lambda}A_\mu\partial_\nu A_\lambda = -16\,\partial_\mu\Big(f(\vec{\zeta})\partial_\nu\zeta^2\partial_\lambda\zeta^3\Big) \qquad (2.99)$$

where:

$$f(\vec{\zeta}) = \frac{\zeta^1}{(1+\vec{\zeta}^2)^3} + 6\int^{\zeta^1}d\zeta'\,\frac{(\zeta')^2 d\zeta'}{(1+\vec{\zeta}'^2)^4}\,;\,\vec{\zeta}'=(\zeta',\zeta^2,\zeta^3) \qquad (2.100)$$

Though the form of $f(\vec{\zeta})$ is not particularly inspiring (nor, of course, unique), (2.99) shows, at a local level, how a topological density can be written as a total divergence.

2.5. THE GINZBURG-LANDAU THEORY OF SUPERCONDUCTIVITY.

The form proposed in the early Fifties by Ginzburg and Landau for the free-energy difference between the superconducting and the normal phase is:

$$\Delta F \equiv F_s - F_n = \int d\vec{x}\ \Delta f(\vec{x}) \qquad (2.101)$$

where:

$$\Delta f = \frac{1}{2\mu}\ |(-i\hbar \nabla + e^*\vec{A}/c)\psi|^2 + \frac{b}{2}(|\psi|^2 - F(T))^2 + \frac{\vec{B}^2}{8\pi}$$
$$(2.102)$$

In (2.102), $\vec{x} \in \mathbb{R}^d$ for some d, $\vec{B} = \nabla \times \vec{A}$ is the magnetic field, μ and e^*
are the mass and charge associated with the complex order parameter ψ, b and F are phenomenological parameters and T is the temperature. It is assumed that

$$b(T) > 0 \ , \quad F(T) = -F_0\ (T - T_c) \ , \quad F_0 > 0 \qquad (2.103)$$

where T_c is the transition temperature.

Eqns. (2.101-102) contain, in a sense, the **minimal** set of ingredients needed to give a description of a superconductor in terms of a complex order-parameter field, namely:

i) Whenever we can set $\vec{A}$ equal to a pure gauge ($\vec{A} = \nabla \lambda$ for some $\lambda = \lambda(x)$), the first term in (2.102) (after an essentially trivial rephasing of ψ) is proportional to $|\nabla \psi|^2$. It is therefore a gradient term which acts to stabilize the uniform solution $\psi =$ const. When $\vec{A}$ is not trivial, it gets modified into a **minimally coupled** gradient term.

ii) The second term is a lowest-order series expansion of the free energy in powers of $|\psi|$. For $\vec{B} = 0$, $\vec{A}$ can be gauged away in any simply-connected region. Stable solutions will correspond to: $|\psi| =$ const., and the free energy has the form depicted in Fig. 15. For $T > T_c$ there is a single minimum at $\psi = 0$, the normal phase, while for $T < T_c$ the free energy exhibits two symmetric minima at:

$$|\psi| = \pm\sqrt{F} \qquad (2.104)$$

iii) The last term is just the free energy density stored in the magnetic field.

iv) The theory is fully gauge-invariant under:

$$\vec{A} \rightarrow \vec{A} - \nabla \Lambda \; ; \; \psi \rightarrow \psi \exp[ie^*\Lambda/\hbar c] \qquad (2.105)$$

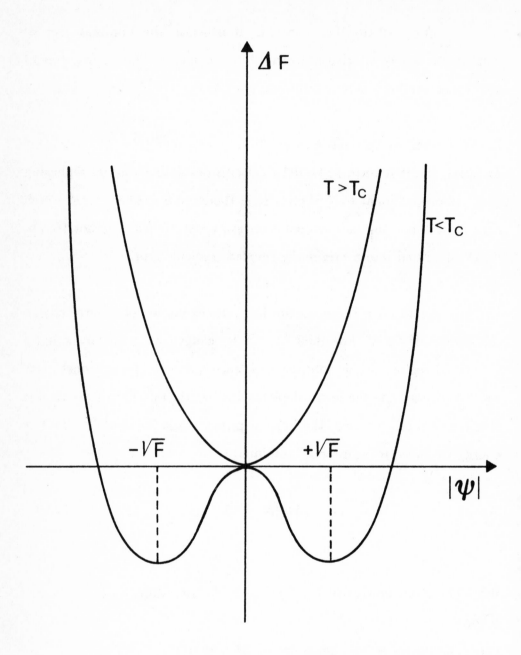

Fig. 15. A qualitative sketch of the free energy in zero field as given by the Ginzburg-Landau theory.

We will concentrate from now on on $T < T_c$. Hence: $F > 0$. Requiring the free energy to be finite requires then:

$$|\psi| \to F , \quad (-i\hbar\nabla + e^*\vec{A}/c)\psi \to 0 \quad \text{for } |\vec{x}| \to \infty \quad (2.106)$$

Setting then: $\psi = |\psi|\exp(i\phi)$, the two conditions (2.106) together imply:

$$\left(\vec{A} - (\hbar c/e^*)\nabla\phi\right) \to 0 \quad \text{for} \quad |\vec{x}| \to \infty \qquad (2.107)$$

It follows from (2.107) that the phase of the order parameter is **not** fixed by the condition that the free energy should be finite. Instead, the boundary conditions at infinity provide us with a map of the surface at infinity onto the set of phases, i.e onto S^1. If $\vec{x} \in \mathbb{R}^d$, we have maps of S^{d-1} onto S^1, and the latter are classified by the homotopy group $\pi_{d-1}(S^1)$. But:

$$\pi_{d-1}(S^1) = 0 \quad \text{for } d \geq 3 \qquad (2.108)$$

Therefore, in general *there are no nontrivial (or stable, in the terminology of Chapt. 1) homotopy sectors for* $d \geq 3$.

The situation changes however rather drastically if d=2, or if d=3 but the superconductor is homogeneous in one direction, in which case the problem is again effectively a two-dimensional one. If this is the case, the relevant homotopy group is $\pi_1(S^1)=\mathbb{Z}$, and we can find topologically nontrivial stable field configurations characterized by winding numbers $n \in \mathbb{Z}$. Integrating then Eq. (2.107) along a large contour $\mathcal{C}$, we obtain:

$$(\hbar c/e^*) \int_\mathcal{C} \nabla\phi \cdot \vec{dl} = \int_\mathcal{C} \vec{A} \cdot \vec{dl} \qquad (2.109)$$

But the r.h.s. is just the total magnetic flux Φ threading the superconductor, while:

$$\int \nabla\phi \cdot \vec{dl} = 2\pi n \qquad (2.110)$$

with n the winding number. We obtain therefore in a simple context the celebrated *flux quantization condition* :

$$\Phi = n \cdot \left(hc/e^*\right), \quad n \in \mathbb{Z} \qquad (2.111)$$

and it is well known that agreement with experiments is obtained if $e^*=2e$, i.e. if the charge of the order parameter is *twice* the electron's charge.

Remark: Suppose the flux is entirely concentrated in a finite region or, for simplicity, into a single flux line. As the phase is not defined along the line, we see that, by consistency, *the order parameter must vanish* at the location of the flux line. Inside a flux tube the material looses then the superconducting properties, and reverts to the normal state.

These considerations lie at the heart of the prediction of the existence of the so-called "intermediate phase" characterized (in tipe-II superconductors) by the partial penetration of the magnetic field inside the sample in the form of tiny flux tubes, or "fluxoids" carrying a quantized flux.

The free energy density (2.102) can be rewritten also as:

$$\Delta f = \frac{1}{2m} \mid (\nabla - ie\vec{A})\psi \mid^2 + \frac{b}{2}(\mid \psi \mid^2 - F)^2 + \frac{\vec{B}^2}{8\pi} \qquad (2.112)$$

where:

$$e = \frac{e^*}{\hbar c} \; ; \quad m = \frac{\mu}{\hbar^2} \qquad (2.113)$$

So, "modulo" a redefinition of the mass and of the coupling constant, Δf looks exactly the energy density for static field configurations of a classical, complex scalar field minimally coupled to an Abelian gauge field and described by the Lagrangian density:

$$\mathcal{L} = \frac{1}{2} (D_\mu \Phi)^*(D^\mu \Phi) - V(\mid \Phi \mid) - \frac{1}{4} F_{\mu\nu} F^{\mu\nu} \qquad (2.114)$$

where: $\Phi = \psi/\sqrt{m}$ and: $D_\mu = \partial_\mu - ieA_\mu$, with an appropriate choice of the "potential" V.

Setting:

$$\psi = \rho \exp[i\phi], \quad \rho = \mid \psi \mid \qquad (2.115)$$

(2.112) becomes:

$$\Delta f = \frac{1}{2m} |\nabla\rho|^2 + \frac{b}{2}(\rho^2 - F)^2 + \frac{\rho^2}{2m} |\nabla\phi - e\vec{A}|^2 + \frac{\vec{B}^2}{8\pi} \quad (2.116)$$

In any **simply connected** region the phase of the order parameter will be single-valued, and can be used then to "regauge" $\vec{A}$. Defining

$$\vec{A}' =: \vec{A} - \frac{1}{e}\nabla\phi \;(\Rightarrow \vec{B}' = \vec{B}) \quad (2.117)$$

we obtain:

$$\Delta f = \frac{1}{2m} |\nabla\rho|^2 + \frac{b}{2}(\rho^2 - F)^2 + \frac{\rho^2 e^2}{2m} \vec{A}'^2 + \frac{\vec{B}^2}{8\pi} \quad (2.118)$$

Varying w.r.t. $\vec{A}$, and using:

$$B_i = \epsilon_{ijk}\partial^i A^k \;; \quad \epsilon^{qmk}B_k = \partial^q A^m - \partial^m A^q \quad (2.119)$$

we find, with some algebra:

$$\delta(\vec{B}^2) = 2\nabla\cdot(\delta\vec{A}\times\vec{B}) - 2\delta\vec{A}\cdot(\nabla^2\vec{A} - \nabla(\nabla\cdot\vec{A})) \quad (2.120)$$

while, of course: $\delta(\vec{A}'^2) = 2\vec{A}'\cdot\delta\vec{A}'$. We find then, for $\vec{A}'$, the field equation:

$$\nabla^2\vec{A}' - \nabla(\nabla\cdot\vec{A}') - \frac{1}{\lambda^2}\vec{A}' = 0 \quad (2.121)$$

where the **penetration depth** λ is given by:

$$\lambda^2 = \frac{2m}{\rho^2 e^2} \equiv \frac{2\mu c^2}{\rho^2 e^{*2}} \tag{2.122}$$

In the **London gauge** defined by $\nabla \cdot \vec{A}' = 0$ Eq.(2.121) becomes:

$$\nabla^2 \vec{A}' - \frac{1}{\lambda^2} \vec{A}' = 0 \tag{2.123}$$

which is the famous **London equation** for the Meissner effect. More generally, taking the curl of Eq. (2.121), we obtain the gauge-invariant form of the London equation, namely:

$$\nabla^2 \vec{B} - \frac{1}{\lambda^2} \vec{B} = 0 \tag{2.124}$$

Considering now the field Lagrangian (2.114), and defining again:

$$\Phi = \rho \, \exp(i\phi) \tag{2.125}$$

we find:

$$\mathcal{L} = \tfrac{1}{2}(\partial_\mu\rho)(\partial^\mu\rho) - V(\rho) + \tfrac{1}{2}\rho^2(\partial_\mu\phi - eA_\mu)(\partial^\mu\phi - eA^\mu) -$$

$$-\tfrac{1}{4}F^{\mu\nu}F_{\mu\nu} \quad (2.126)$$

If the coupling constant vanishes (e → 0), the Lagrangian becomes:

$$\mathcal{L}_0 = \tfrac{1}{2}(\partial_\mu\rho)(\partial^\mu\rho) - V(\rho) - \tfrac{1}{4}F_{\mu\nu}F^{\mu\nu} + \tfrac{1}{2}\rho^2(\partial_\mu\phi)(\partial^\mu\phi) \quad (2.127)$$

If $\rho \neq 0$, the phase of the field behaves as an independent **massless** field. It exists only when $\rho \neq 0$, i.e. for field solutions which break the $\cup(1)$ gauge invariance of the theory. This phenomenon of **spontaneous symmetry breaking** brings about the existence of a massless **Goldstone mode** [84] in the theory.

If instead e ≠ 0, and in a simply-connected region, we can again regauge A_μ as:

$$A_\mu \to A'_\mu = A_\mu + \tfrac{1}{e}\partial_\mu\phi \quad (2.128)$$

and the Lagrangian becomes:

$$\pounds = \tfrac{1}{2}\,(\partial_\mu\rho)(\partial^\mu\rho) - V(\rho) + \frac{e^2\rho^2}{2}\,A'_\mu A'^\mu - \tfrac{1}{4}\,F_{\mu\nu}F^{\mu\nu} \qquad (2.129)$$

The effect of the coupling with the gauge field is that *the Goldstone mode disappears, and the gauge field acquires a mass term.* This is known as the *Higgs phenomenon* [84] in Field Theory, and is exactly what we have seen to happen in the context of the Ginzburg-Landau theory. It implies, e.g., exponential decay of the magnetic field inside a slab of superconducting material on the scale of the penetration depth (1.122), and this is well known as the Meissner effect *The Meissner effect is therefore a macroscopic manifestation of the Higgs phenomenon.*

CH.3. INEQUIVALENT QUANTIZATIONS IN MULTIPLY CONNECTED SPACES. BRAID GROUPS AND ANYONS.

3.1. INTRODUCTION.

In this Chapter we will be concerned mainly with **Quantum Mechanics in a nonrelativistic context and in the Schrödinger picture.** Moreover, we will consider only **scalar** Quantum Mechanics. All in all, this amounts to assume that the physical systems we will be concerned with are described by complex wave functions:

$$\Psi \in \mathbb{L}_2(\mathbb{Q},\mathbb{C},\mathrm{d}\mu) \tag{3.1}$$

where $\mathbb{Q}$ is the system's configuration space and $\mathrm{d}\mu$ is a measure on $\mathbb{Q}$ (typically the Lebesgue measure).

The time evolution of the system is governed by the Schrödinger equation:

$$i\hbar\partial_t\Psi = \hat{\mathsf{H}}\ \Psi \tag{3.2}$$

where $\hat{\mathsf{H}}$ is the Hamiltonian. $\hat{\mathsf{H}}$ (as well as any other observable pertaining to the system) is a *local* operator (typically a differential operator) which is, to start with, a *symmetric* operator that can be extended to a *self-adjoint* operator once its domain is properly and carefully defined. In particular, this requires that the boundary conditions of Ψ on Q be carefully specified. *The boundary conditions become then an essential "ingredient" in the definition of the domain (if any) of self-adjointness of* $\hat{H}$.

Let me stress here an almost obvious fact, namely that, although Ψ is complex, *all of the physical information is contained in the squared modulus* $|\Psi(q)|^2$ $(q \in Q)$ *of* Ψ. In other words, we can freely modify Ψ by a $\mathbb{U}(1)$ phase factor:

$$\Psi(q) \rightarrow \Psi(q) \cdot \exp[i\alpha(q)] \; ; \; \alpha \in \mathcal{F}(Q) \tag{3.3}$$

without altering the physical description[1]. It appears therefore that a complex wave function, though it is unavoidable in general, offers

----

(1). *This may require modifying the differential operators like* ∇ *appearing in* $\mathbb{H}$ *into* **covariant derivatives***, i.e.:*

$$-i\hbar\nabla \mapsto -i\hbar\nabla - \vec{A}$$

where, under (3.3):

$$\vec{A} \mapsto \vec{A} + \nabla\alpha/\hbar$$

nonetheless a **redundant** description of the status of affairs, as an overall phase (of course *not* a *relative* phase in the linear superposition of wave functions!) is unobservable.

Whether or not the phase of a wave function can be fixed *globally* as a well-defined $U(1)$-valued function on the whole of Q has been discussed since the early days of Quantum Mechanics (see,e.g., the papers of W. Pauli quoted in [56]) with the main conclusion that *as long as Q is simply connected, a global definition of the phase is always possible.*

Digression: We recall [30] that a principal fiber bundle admits of a global section iff it is a trivial bundle (i.e. a direct product bundle). It seems therefore that, if Q is simply connected, we can consider a trivial bundle of the form $Q \times C$, and that the wave function is a global section of such a bundle. The latter is in turn an associated bundle [30] of the (trivial) principal bundle $Q \times U(1)$.

However, when Q is not simply connected, there may arise ambiguities in the definition of the phase of the wave function, and *it may not be always possible to define the latter globally* as a well-defined (i.e. single-valued and $U(1)$-valued) function on Q. In this Chapter we will address to the problems arising from the ambiguity in the global definition of the wave function.

3.2. QUANTUM MECHANICS IN NONSIMPLY CONNECTED SPACES.

Let then the configuration space Q be nonsimply connected, i.e. $\pi_1(Q) \neq 0$.

Let's recall that for any topological space Q we can construct the (unique up to isomorphisms) **universal covering space** $\tilde{Q}$ [90] with the following properties:

i) $\tilde{Q}$ is simply connected: $\pi_1(\tilde{Q}) = 0$.

ii) There is a *covering projection:*

$$\pi: \tilde{Q} \rightarrow Q \qquad (3.4)$$

iii) There is a *free* (i.e. transitive and effective) action of $\pi_1(Q)$ on $\tilde{Q}$ s.t.:

iv) The action permutes the "preimages" of any $q \in Q$ (i.e. the points $\tilde{q} \in \pi^{-1}(q) \subset \tilde{Q}$) among themselves, and:

v) There is an open covering of Q s.t. for any set $\mathfrak{U}$ in the covering $\pi^{-1}(\mathfrak{U})$ is a *disjoint union* of open sets in $\tilde{Q}$, and the restriction of π to any one of the latter turns out to be an *isomorphism*.

Remark: If Q is a differentiable manifold, it can be proved [90] (though we will not do it here) that $\tilde{Q}$ can be given a differentiable structure s.t. π in v) becomes actually a *diffeomorphism*.

The Hamiltonian (as well as the other observables), being by assumption a **local** operator, can be "lifted" without ambiguities to (or redefined as) a **symmetric operator** on $\tilde{Q}$. We can then construct Quantum Mechanics on the universal covering space in a nonambiguous way with single-valued wave functions, globally defined phases and so on. The relevant question becomes then:

Under which conditions will the Quantum Mechanics on $\tilde{Q}$ be "projectable" to a (sensible) Quantum Mechanics on Q ?

Otherwise stated, the question is:

What kind of boundary conditions should we impose on wave functions on $\tilde{Q}$ (and hence which domains of self-adjointness on the Hamiltonian and on the other relevant observables) in such a way that, calling $\tilde{\Psi} \in L_2(\tilde{Q},C,d\mu)$ a wave function on $\tilde{Q}$, $|\tilde{\Psi}(\tilde{q})|^2$, $\tilde{q} \in \tilde{Q}$, be projectable, i.e. depends only on $q=\pi(\tilde{q})$?

As a preliminary, let's recall the following

Theorem :

Let $q \in Q$, consider a loop γ at q, the corresponding homotopy class $[\gamma] \in \pi_1(Q,q)$, and an arbitrary preimage $\tilde{q}$ of q in $\tilde{Q}$. Then γ lifts to a curve $\tilde{\gamma}$ in $\tilde{Q}$ beginning at $\tilde{q}$ and ending at the point $\tilde{q}'$ which is obtained from $\tilde{q}$ by the action of $[\gamma]$. In symbols:

$$\tilde{q}' = [\gamma] \cdot \tilde{q} \qquad (3.5)$$

We refer to the literature [57,90,94] for the proof of this "lifting homotopy theorem". $\square$

Remark: It is only homotopically trivial loops in Q that lift to loops in $\tilde{Q}$, and viceversa. Also, any two curves joining $\tilde{q}$ and $\tilde{q}'$ in $\tilde{Q}$ will project down to loops in $[\gamma]$ (and viceversa again).

If we want $|\tilde{\Psi}(\tilde{q})|^2$ to be projectable, we must require that $\tilde{\Psi}(\tilde{q})$ and $\vec{\Psi}(\tilde{q}')$ *differ at most by a phase* for any $\tilde{q}$, $\tilde{q}' \in \pi^{-1}(q)$. On the other hand, in view of the single-valuedness of $\tilde{\Psi}$, the phase *cannot* depend on the path that has been followed to go from $\tilde{q}$ to $\tilde{q}'$. *It can depend only on the group element* $[\gamma]$ *that connects* $\tilde{q}$ *and* $\tilde{q}'$. We then have our central result, namely [12-15,37,56,58,59,65,67,87,93,96]:

Theorem: *Projectable Quantum Mechanics are obtained iff the wave functions on the universal covering space obey the boundary conditions:*

$$\tilde{\Psi}([\gamma] \cdot \tilde{q}) = a([\gamma]) \, \tilde{\Psi}(\tilde{q}) \quad \forall \, \tilde{q} \in \tilde{Q} \qquad (3.6)$$

where: $|a([\gamma])| = 1 \, \forall \, [\gamma] \in \pi_1(Q)$. $\square$

Further restrictions on the admissible phases $a([\gamma])$ follow again from the single-valuedness of $\tilde{\Psi}$. Consider three points $\tilde{q}$, $\tilde{q}'$ and $\tilde{q}''$ s.t.:

$$\tilde{q}' = [\gamma] \cdot \tilde{q} \; ; \;\; \tilde{q}'' = [\gamma'] \cdot \tilde{q}' \;\; \Rightarrow \tilde{q}'' = [\gamma] \cdot [\gamma'] \cdot \tilde{q} \equiv [\gamma \cdot \gamma'] \cdot \tilde{q} \quad (3.7)$$

Single-valuedness of $\tilde{\Psi}$ implies then:

$$a([\gamma]) \cdot a([\gamma']) = a([\gamma \cdot \gamma']) \equiv a([\gamma] \cdot [\gamma']) \quad (3.8)$$

We have then proved the following:

Theorem: *The map:*

$$a \colon \pi_1(Q) \to U(1) \quad \text{by:} \quad [\gamma] \to a([\gamma]) \quad (3.9)$$

is a one-dimensional unitary representation of $\pi_1(Q)$. $\square$.

We recall that the **commutator subgroup** of a group G is the minimum subgroup containing all the elements of the form $g \cdot h \cdot g^{-1} \cdot h^{-1}$, for $g, h \in G$. It is denoted by $[\![G, G]\!]$. Taking then two elements $[\gamma]$ and $[\mu]$ in $\pi_1(Q)$, we get from the previous theorem:

$$a([\gamma]^{-1}) = a([\gamma])^* \; ; \;\; a([\mu]^{-1}) = a([\mu])^* \quad (3.10)$$

and:

$$a([\gamma][\mu][\gamma]^{-1}[\mu]^{-1})= a([\gamma])a([\mu])a([\gamma])^*a([\mu])^*= 1 \qquad (3.11)$$

It follows then that the map (3.9) acts as the identity on the commutator subgroup $[\![\pi_1,\pi_1]\!]$.

It is known [54,57] that the quotient of π_1 w.r.t. $[\![\pi_1,\pi_1]\!]$, also called the "**abelianization**" of π_1, is the *first homology group with integer coefficients*, an abelian group denoted by $\mathbb{H}_1(\mathbb{Q},\mathbb{Z})$:

$$\mathbb{H}_1(\mathbb{Q},\mathbb{Z})= \pi_1/[\![\pi_1,\pi_1]\!] \qquad (3.12)$$

Then, the map (3.9) is actually a one-dimensional representation of $\mathbb{H}_1$. It is also known as a *character* of the original group π_1.

The next problem concerns what we can tell about the nature of the wave function $\Psi(q)$ on the original configuration space $\mathbb{Q}$.

It is shown [90] in the theory of covering spaces that $\tilde{\mathbb{Q}}$ can be decomposed into the union of a set of "**fundamental domains**", each one of which is isomorphic (diffeomorphic if we consider differentiable structures) to $\mathbb{Q}$. Each fundamental domain contains one (and only one) preimage of every point $q \in \mathbb{Q}$, and different fundamental domains are connected by the action of π_1 on $\tilde{\mathbb{Q}}$. The wave function $\tilde{\Psi}$ on a fundamental domain $\mathbb{F} \in \tilde{\mathbb{Q}}$ projects then down to a well defined wave function $\Psi(q)$ on $\mathbb{Q}$. However, if we take the point q around a loop (based at q), at the end of the process we end up with the projection of the wave function coming from a different fundamental domain.

In view of Eq. (3.6), we conclude that:

To the admissible Quantum Mechanics on the universal covering space there correspond in general <u>multivalued</u> wave functions on **Q**. *If* Ψ(q) *is one such wave function, Quantum Mechanics on* Q̃ *defines an <u>action</u> of* π₁ *on* Ψ, *whereby taking the point q around a loop* γ *in* **Q** *multiplies the wavefunction by the phase a([γ]).*

In conclusion, we can say that *there is nothing wrong* (as far as the physical interpretation of Quantum Mechanics goes) *in the use of multivalued wave functions, provided the configuration space is nonsimply connected, and that the admissible multivaluednesses are limited to multiplication by characters of the fundamental group of the configuration space.*

Consider now two wavefunctions Ψ_1 and Ψ_2, and assume that, by going around a loop γ:

$$\Psi_i(q) \rightarrow a_i([\gamma]) \cdot \Psi_i(q) \ , \ i= 1,2 \ ; \tag{3.13}$$

If we assume that the superposition principle holds unrestricted, we can form the wave function:

$$\Psi = \Psi_1 + \Psi_2 \tag{3.14}$$

But then circling around the loop γ yields:

$$\Psi \rightarrow a_1 \Big\{ \Psi_1 + a_1^* a_2 \Psi_2 \Big\} \tag{3.15}$$

Therefor, unless $a_1 = a_2$, we have produced a **relative** phase $a_1^* a_2$ which can of course lead to observable interference effects. Otherwise stated, $|\Psi|^2$ **is not projectable.** It follows that *the superposition principle can hold only among wave functions which undergo multiplication by the* <u>*same*</u> *character when taken around a loop in* **Q**. This restriction has the typical form of a *superselection rule* [56], and we conclude that:

> *If* **Q** *is nonsimply connected, the Hilbert space of states* ($L_2(Q,C,d\mu)$ *in the Schrödinger picture) decomposes into a direct sum of* <u>*superselection*</u> <u>*sectors*</u>, *each sector being characterized by a given character* **a** *multiplying all wave functions in the sector by* $a([\gamma])$ *when they are taken around a loop* γ *in* **Q**.

Digression: Let's go back for a moment to $\tilde{Q}$ and to the boundary condition (3.6). It is amusing to note [87] that there is a simple physical system that realizes this situation. Consider (for simplicity) a one-dimensional lattice with lattice spacing d. Electrons moving in a potential having the periodicity of the lattice are described by **Bloch functions** $\psi_{nk}(x)$, with n a band index and k a (quasi)momentum index. The Bloch functions satisfy the boundary condition:

$$\psi_{nk}(x+d)= \exp[ikd]\psi_{nk}(x) \qquad (3.16)$$

But then we can view the unit cell, with endpoints identified, as S^1, whose universal covering is $\mathbb{R}$. As $\pi_1(S^1)=\mathbb{Z}$, the characters of $\mathbb{Z}$ are exactly of the form $\exp[ikd]$, $|k| \leq \pi/d$, and are labeled by the number kd. Electrons on a **perfect** lattice provide then us with a concrete realization of Quantum Mechanics on a nonsimply connected space and on its universal covering. For this reason, we will call the boundary conditions (3.6) *"generalized Bloch conditions"*.

In what follows we will concentrate on the case of identical particles in $\mathbb{R}^3$ and $\mathbb{R}^2$. We refer to the literature [12-15,24,37,56,58,59,71,87,93,96] for the discussion of other examples of Quantum Mechanics in nonsimply connected spaces.

3.3. THE CASE OF IDENTICAL PARTICLES.

i) Identification of the proper configuration space.

We consider from now on identical particles moving in $\mathbb{R}^d$ for some $d \geq 2$. Later on we will specialize to the case d=2. For N identical particles, we start obviously from the N-fold Cartesian product:

$$X_N= \mathbb{R}^d \times ... \times \mathbb{R}^d= \mathbb{R}^{Nd} \qquad (3.17)$$

as a candidate for the configuration space. However, as the particles are indistinguishable, we have to make some identifications in X_N. Actually, we have to identify any two configurations which differ only by the exchange of two particles. Exchange can be defined through an action of the symmetric group S_N on X_N. As a further step, we consider then as a candidate for the configuration space:

$$Q_0 = X_N / S_N \qquad (3.18)$$

i.e. the quotient of X_N by the action of S_N. This leads however to some technical difficulties. Indeed, let $\vec{x}_1....\vec{x}_N$ be the coordinates of the particles, and let $\sigma_{i,j}$ denote the operation of interchange of $\vec{x}_i$ and $\vec{x}_j$. If the particles happen to coincide in space ($\vec{x}_i = \vec{x}_j$), $\sigma_{i,j}$ will leave any such configuration unaltered. In other words, the action of S_N on X_N may have **fixed points**. The locus of the latter is the **diagonal** Δ of X_N, defined as:

$$\Delta = \left\{ \vec{x}_1....\vec{x}_N; \ \vec{x}_i \in \mathbb{R}^d \vdash \vec{x}_i = \vec{x}_j \text{ for at least a pair (i,j), } i \neq j \right\}$$
$$(3.19)$$

Now, it so happens that any time the action of a group on a manifold has fixed points, the quotient w.r.t. this action can be a manifold with some pathologies (e.g. it can "develop" boundaries) or can even be no manifold at all [24].

Exercise: Let d=1, N=2. Prove that X_2/S_2 is a manifold with boundary (a half plane).

In order to avoid pathologies, we resort to defining *the configuration space of N identical particles moving in* $\mathbf{R}^d$ *as:*

$$Q= (\mathbf{R}^{Nd} - \Delta)/S_N \tag{3.20}$$

Remark: The motivations that we gave for excluding the diagonal were eminently technical. Various authors [106,107] have argued that this exclusion is physically harmless.

ii) What is $\pi_1(Q)$.

Let's recall that the permutation group S_N is a finite group (of order N!) which can be given a finite "presentation" [85] in terms of N abstract generators $\sigma_1....\sigma_{N-1}$ (plus the identity) satisfying the relations listed below:

$$S_N = \ <e,\sigma_1...\sigma_{N-1} \mid \sigma_i\sigma_{i+1}\sigma_i=\sigma_{i+1}\sigma_i\sigma_{i+1} \ ; \ \sigma_i\sigma_j=\sigma_j\sigma_i, \ |i-j| \geq 2 \ ; \ \sigma_i^2=e>$$

$$\tag{3.21}$$

If σ_i is seen concretely as the operation of interchange of the i-th and (i+1)-th objects in an assembly of N objects, all the relations in (3.21) have a transparent meaning.

We will need in a short while another group, the so-called **N-*string braid group*, $\mathbb{B}_N$** [9,23,50]. It is an **infinite** discrete group that can be defined again in terms of the identity and of N-1 generator $\sigma_1....\sigma_{N-1}$, and has the following finite presentation:

$$\mathbb{B}_N= <e,\sigma_1...\sigma_N \mid \sigma_i\sigma_{i+1}\sigma_i= \sigma_{i+1}\sigma_i\sigma_{i+1} ; \sigma_i\sigma_j= \sigma_j\sigma_i, |i\text{-}j| \geq 2 > \tag{3.22}$$

(3.22) has been obtained from (3.21) simply by deleting the condition $\sigma_i^2=e$. However, $\mathbb{B}_N$ is profoundly different from S_N. In the first place it is an infinite group (each element is of infinite order). Also, the characters of the group are different, namely:

i)   The characters of S_N can be generated by associating a phase of $\exp[-i\theta_j]$ to the generator σ_j. The property $\sigma_i\sigma_j=\sigma_j\sigma_i$ for $|i\text{-}j| \geq 2$ is trivially·satisfied. $\sigma_i\sigma_{i+1}\sigma_i=\sigma_{i+1}\sigma_i\sigma_{i+1}$ leads in turn to:

$$\theta_j= \text{(independent of j)}= \theta \tag{3.23}$$

while, finally, $\sigma_i^2= e$ leads to

$$\theta= 0 \text{ or } \pi \pmod{2\pi} \tag{3.24}$$

Therefore, apart from $a(e)= 1$, we can only have either $a(\sigma_i)= 1 \; \forall i$ or $a(\sigma_i)= -1 \; \forall i$. If g is a generic element of S_N, we have then the two possibilities:

$$a(g) = 1 \ \forall \ g \qquad\qquad (3.25\text{-a})$$

or

$$a(g) = (-1)^g \qquad\qquad (3.25\text{-b})$$

where $(-1)^g$ is $+1$ for even permutations, and -1 for odd ones.

ii) As far as $\mathbb{B}_N$ is concerned, we don't have the condition $\sigma_i^2 = e$, so (cfr. Eq. (3.23)):

$$a(e) = 1 \ ; \ \ a(\sigma_j) = \exp[-i\theta] \ , \ \ j = 1,..., N-1 \qquad\qquad (3.26)$$

The characters of the braid groups are therefore labeled by an angle θ which can take any value between 0 and 2π: $0 \leq \theta < 2\pi$.

Now, the following theorem holds:

Theorem: *With $\mathbf{Q}$ given by Eq. (3.20), we have:*

$$\pi_1(\mathbf{Q}) = \mathbf{S_N} \ \textit{for} \ \ d \geq 3 \qquad\qquad (3.27\text{-a})$$

$$\pi_1(\mathbf{Q}) = \mathbf{B_N} \ \ \textit{for} \ d = 2 \qquad\qquad (3.27\text{-b})$$

I will not give a full proof of the theorem here, but wiil rather give some intuitive justifications of it. Consider first the case d=3 (what we will say for d=3 will hold "a fortiori" for d > 3). An elementary loop in **Q** corresponds to the interchange of the positions of two particles which I will label as "i" and "i+1". We are thus considering the action of σ_i. Taking a plane through the positions of the particles, we can depict the operation of interchange and its inverse as in Fig. 16.

It is easy to see that, if $|i-j| \geq 2$, σ_i and σ_j are independent operations and that the final result does not depend on the order in which the two operations have been performed. So: $\sigma_i \sigma_j = \sigma_j \sigma_i$ for $|i-j| \geq 2$. That $\sigma_i \sigma_{i+1} \sigma_i = \sigma_{i+1} \sigma_i \sigma_{i+1}$ can be proved in a similar way. Indeed, let's indicate by $\odot$, $\otimes$ and $\ominus$ the particles labeled by i, i+1 an i+2 respectively. Then:

$$\sigma_i \sigma_{i+1} \sigma_i = (\odot, \otimes, \ominus) \xrightarrow{\sigma_i} (\otimes, \odot, \ominus) \xrightarrow{\sigma_{i+1}} (\otimes, \ominus, \odot) \xrightarrow{\sigma_i} (\ominus, \otimes, \odot)$$

$$(3.28\text{-a})$$

while:

$$\sigma_{i+1} \sigma_i \sigma_{i+1} = (\odot, \otimes, \ominus) \xrightarrow{\sigma_{i+1}} (\odot, \ominus, \otimes) \xrightarrow{\sigma_i} (\ominus, \odot, \otimes) \xrightarrow{\sigma_{i+1}} (\ominus, \otimes, \odot)$$

$$(3.28\text{-b})$$

which proves the asserted equality.□.

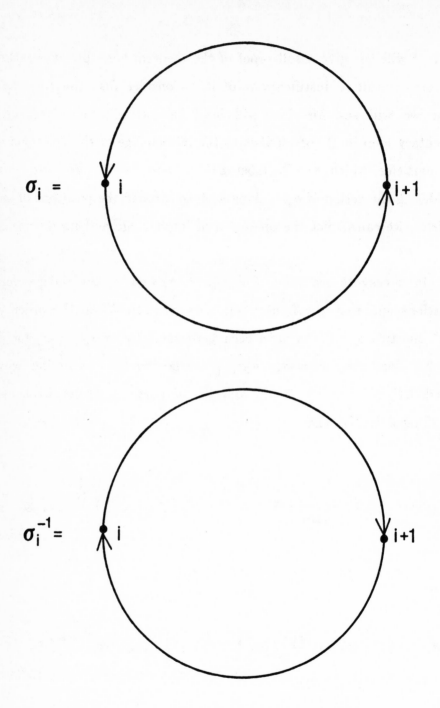

Fig. 16. The loops in $\mathbf{Q} = \mathbb{R}^2$ corresponding to σ_i and σ_i^{-1}.

Let's take advantage now of the fact that our particles "live" in a three-dimensional world. Consider the plane of Fig. 16 as, say, the x-y plane. By a continuous rotation in space around the axis joining the particles' positions, *we can homotopically deform σ_i into σ_i^{-1}*. Hence:

$$\sigma_i \sim \sigma_i^{-1} \Rightarrow \sigma_i^2 \sim e \qquad (3.29)$$

which completes the (sketch of the) proof that $\pi_1(\mathbb{Q}) = S_N$ for d=3 (and also for d > 3).

Rotations in the third dimension is precisely what we are not allowed to perform when d=2. In this case, therefore, σ_i and σ_i^{-1} are *not* homotopic, and $\sigma_i^2 \sim e$ doesn't hold anymore. It is the absence of this relation that makes the difference between S_N and $\mathbb{B}_N$, so we conclude that $\pi_1(\mathbb{Q}) = \mathbb{B}_N$ for d= 2.□.

Eqns. (3.27) have profound consequences. For d $\geq$ 3 we see that, in the notation of Eq. (3.13), and if [g] represents (the homotopy class of) any permutation, the wave function changes as:

$$\Psi \rightarrow a([g])\Psi = \begin{cases} \Psi \text{ if } a \equiv 1 \text{ (Bose statistics)} \\ \\ \pm \Psi \text{ if } a = \pm 1 \text{ (Fermi statistics)} \end{cases} \qquad (3.30)$$

Hence, **only Bose or Fermi statistics are allowed in space dimension d ≥ 3** [37,67]. Instead, for d=2:

$$\Psi \rightarrow a([g])\Psi \;\; ; \;\; a([g]) = \exp\left(i \sum (\pm \theta)\right) \tag{3.31}$$

with a $-\theta$ $(+\theta)$ appearing in the sum for each factor of σ_i (σ_i^{-1}) out of whose product $g \in \mathbb{B}_N$ is composed. The limiting cases $\theta=0$, π correspond again to Bose and Fermi statistics, but, in general, the statistics can interpolate continuously between the two limiting cases. Wilczek [105] has coined the term "anyons" for particles obeying (3.31) for generic θ. Also, and to some extent improperly, the term "fractional statistics" is often employed to denote such particles.

Y.S.Wu [111,112] has developed a very interesting interpretation of the "braid statistics" (3.31).

Consider the "unfolding" in time of the process of interchanging the positions of two particles (Fig. 17). For every interchange of the particles labeled by the indices i and j, the relative angle between them (represented by the oriented arrow in Fig. 17) changes by π, or $-\pi$ for the reverse exchange. Then Eq. (3.31) can be rewritten as:

$$a([g]) = \exp\left(-i \frac{\theta}{\pi} \sum_{i < j} \Delta\phi_{ij}\right) \tag{3.32}$$

where $\Delta\phi_{ij}$ is the total variation of the relative angle between particles i and j, and is a (positive or negative) multiple of π. As the process by

133

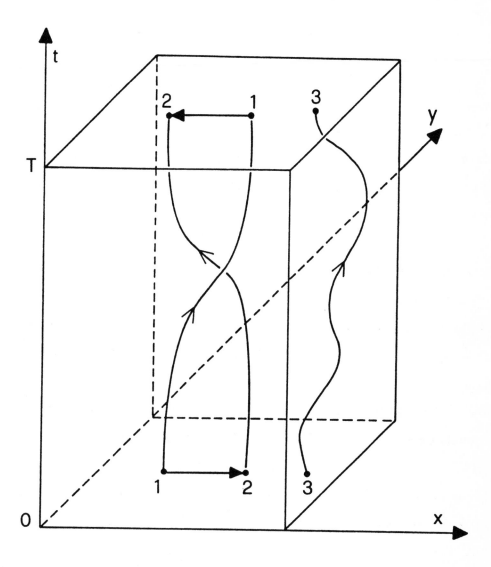

Fig. 17. Unfolding in time of a process of interchange (1⇔2 for N=3).

which ϕ_{ij} changes is smooth in time, we can write also:

$$a([g]) = \exp\left(-i\,\frac{\theta}{\pi}\int\limits_{0}^{T} dt\,\frac{d}{dt}\sum_{i<j}\phi_{ij}(t)\right) \tag{3.33}$$

where the way the ϕ_{ij}'s depend on time depends on the structure of the homotopy class [g], and T is the total time during which the process takes place.

Remarks: i) In the case of, say, Fermi statistics $(\theta=\pi)$, any $\Delta\phi_{ij}$ which is an **even** multiple of π is irrelevant in (3.32). The final result a$= \pm 1$ is not altered by changing the "history" of the pair of particles by a total relative angle of $2k\pi$ for some integer k. This is even more true for Bosons $(\theta=0)$. However, for generic θ, *the phase factor keeps track of the whole history*, and loops differing by an even number of interchanges of i and j will lead to *different* phase factors.

ii) It is not difficult to convince oneself that the (discrete) operation P(the parity which, in two dimensions, amounts to reflection about **one** of the coordinate axes) leads to a reversal of the direction of winding of any two particles around each other. In turn, T (the time reversal) implies [73] a complex conjugation of the wave functions. It follows that the operation of either P or T amounts to changing θ into $-\theta$ in (3.31-33). Therefore, unless $\theta=0$ or $\theta=\pi$ $(=-\pi$ mod. $2\pi)$, i.e. unless the particles obey Bose or Fermi statistics, *"fractional" statistics leads to* *spontaneous violation of the discrete symmetries* **P** and **T**, while the combined symmetry **P** · **T** is unbroken.

3.4. DYNAMICAL IMPLEMENTATION OF BRAID STATISTICS.

The fact that in the generic (anyon) case the final phase of the wave function depends on the previous history and not only on the initial and final configurations suggests that we cast the dynamical description of our system in the framework of Feynman's formulation of Quantum Mechanics [37,41,56,65,87] in terms of "path integrals" or "sums over histories".

For a simply connected space, the transition amplitude from position q' at t=0 to position q at time T, with q, q' denoting the whole set of coordinates of the particles (q= $(\vec{x}_1....\vec{x}_N)$ etc.):

$$\mathcal{K}(q,q';T)=: \; <q,T \mid q',0> \qquad (3.34)$$

is written, in the path integral formulation, as:

$$\mathcal{K}(q,q';T)= \int \mathcal{D}\gamma(t) \, \exp\left(iS[\gamma]/\hbar\right) \qquad (3.35)$$

where the symbol "$\mathcal{D}\gamma$" stands for the "sum over histories" [41] and S is the classical action evaluated along a path γ leading from q' to q:

$$S[\gamma]= \int_0^T \mathcal{L}(\gamma(t), \dot{\gamma}(t)) \, dt \; ; \; \gamma(0)= q', \; \gamma(T)= q \qquad (3.36)$$

Finally, $\mathcal{L}$ is the classical Lagrangian. For a simply connected space, Eq.(3.35) defines the transition amplitude in a non ambiguous way, except for an overall phase factor of the form exp $[i(h(q)-h(q'))/\hbar]$, which is however unobservable [56].

Consider now the case in which the configuration space $\mathbf{Q}$ is nonsimply connected. If we choose at will a "preimage" $\tilde{q}'$ of q' on the universal covering $\tilde{\mathbf{Q}}$, we can use again (3.35) to construct Quantum Mechanics on $\tilde{\mathbf{Q}}$. However, all the preimages $\tilde{q}$ of q are now on an equal footing, and we have as many propagators as there are distinct elements in $\pi_1(\mathbf{Q})$. Projecting them down to $\mathbf{Q}$ we see that any two of them correspond to having taken q around a loop belonging to the element of $\pi_1(\mathbf{Q})$ connecting the preimages.

It is easy to prove [56] that, in $\mathbf{Q}$, we obtain a propagator consistent with the multivaluedness condition expressed by Eq. (3.13) if we write $\mathcal{K}$ as:

$$\mathcal{K}(q,q';T)= \sum_{[\alpha] \, \in \, \pi_1(\mathbf{Q})} a([\alpha]) \, \mathcal{K}_\alpha(q,q';T) \qquad (3.37)$$

where $\mathcal{K}_\alpha$ is the propagator (actually, a **partial** propagator) obtained by restricting the path integral to paths belonging to the homotopy class of paths from q' to q which are labeled by the element $[\alpha] \in \pi_1(\mathbf{Q})$.

Note that we are talking here about **open** paths. Therefore, we have to specify how to label them with homotopy classes. With reference to Fig. 18, the assignment of a homotopy class to a path like γ can be achieved once and for all by choosing a fiducial point $q_0 \in \mathbf{Q}$ and by fixing a "homotopy mesh" [67], i.e. a mesh of paths joining q_0 to any other point in $\mathbf{Q}$. In this way we can assign a loop to each path, though in a noncanonical way, and homotopy considerations apply. It can be shown (we refer to the literature [37,67] for a proof of this point) that changing the fiducial point and/or the homotopy mesh alters the propagator by an **overall** phase factor, which is unobservable, as already discussed.

If we take now q around a loop β^{-1}, we find [56] that:

$$\mathcal{K}_\alpha(q,q';T) \rightarrow \mathcal{K}_{\alpha\beta^{-1}}(q,q';T) \tag{3.38}$$

Hence:

$$\mathcal{K} \rightarrow \sum_{[\alpha]} a([\alpha \cdot \beta^{-1}])\, a([\beta])\, \mathcal{K}_{\alpha\beta^{-1}} \equiv a([\beta])\, \mathcal{K} \tag{3.39}$$

and this achieves the proof that (3.37) is consistent with the required multivaluedness of the wave functions.$\square$

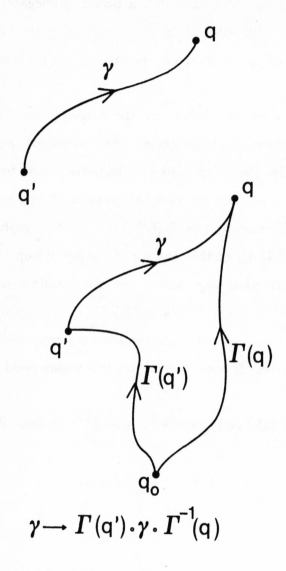

Fig.18. Construction of the "homotopy mesh" discussed in Sect. 3.4.

Let's put now Eqns. (3.33) and (3.37) together. With reference to Fig. 18, we see that we can rewrite the character a([α]) in (3.37) as:

$$a([\alpha]) = \exp\left(-i\frac{\theta}{\pi}\sum_{i<j}\int_{\Gamma(q)}d\phi_{ij}\right)\cdot\exp\left(-i\frac{\theta}{\pi}\sum_{i<j}\int_{\Gamma(q')}d\phi_{ij}\right)\times$$

$$\times\exp\left(-i\frac{\theta}{\pi}\sum_{i<j}\int_{\gamma}d\phi_{ij}\right) \equiv \qquad (3.40)$$

$$\equiv \chi_0(q)\chi_0(q')^*\exp\left(-i\frac{\theta}{\pi}\sum_{ij}\int_{\gamma}d\phi_{ij}\right)$$

where:

$$\chi_0(q) = \exp\left(-i\frac{\theta}{\pi}\sum_{ij}\int_{\Gamma(q)}d\phi_{ij}\right) \qquad (3.41)$$

and $\chi_0(q')$ are defined relative to the homotopy mesh of Fig. 18, and the path γ is labeled by the element $[\alpha]\in\pi_1(Q)$ in the sense already specified above.

The last factor in the last line of (3.40) can be reabsorbed in the propagator. The sum over homotopy classes can be performed, and the final form of the propagator reads:

$$\mathcal{K}(q,q';T) = \chi_0(q)\chi_0(q')^*\int\mathcal{D}\gamma(t)\,\exp\left[i\tilde{S}[\gamma]/\hbar\right] \equiv$$

$$\qquad (3.42)$$

$$\equiv \chi_0(q)\chi_0(q')^*\mathcal{K}_0(q,q';T)$$

where the integral is now over the whole space of paths,

$$\tilde{S}[\gamma] = \int_0^T dt\ \tilde{\mathcal{L}}(\gamma(t),\ \dot\gamma(t)) \tag{3.43}$$

and the modified Lagrangian $\tilde{\mathcal{L}}$ is given by:

$$\tilde{\mathcal{L}} = \mathcal{L} - \frac{\hbar\theta}{\pi} \sum_{i<j} \frac{d}{dt} \phi_{ij} \tag{3.44}$$

The phase factors $\chi_0(q)$ and $\chi_0(q')$ do not affect the absolute value of the propagator. Hence, the physics of the system is entirely contained in the "reduced" propagator $\mathcal{K}_0(q,q';T)$.

A gauge transformation, i.e. the addition to the Lagrangian of the total time derivative of a function $h(q)$, would modify the propagator by a similar product of phase factors, that is by $\exp[i(h(q) - h(q'))/\hbar]$. However, the χ_0's are *path-dependent* phase factors. As such, they cannot be disposed of (or gauged away) by performing a (globally defined) gauge transformation. However, as they do not affect the physics, they can be omitted and the system can be described by the propagator $\mathcal{K}_0(q,q';T)$ defined in (3.42). This is what in the literature ([106,107] and references therein] has been termed as a "singular gauge transformation". Note that $\mathcal{K}_0$ is *single-valued* [56,76].

However, it contains an additional interaction term in the effective Lagrangian, namely the last term in (3.44).The latter is *locally* a total time derivative. As such, it does not affect the **classical** equations of motion, but it has profound effects at the **quantum** level.

As the wave functions are now single-valued, what we are actually achieving by means of a singular gauge transformation is an equivalent description of the system in the "trivial" superselection sector, at the price however of introducing in the Lagrangian an additional, statistical, interaction. Anyons can then be described either as particles with "exotic" statistics determined by the mutivaluedness of their wave functions, or as "ordinary" particles, which however interact in a manner that is entirely quantum in nature and has no classical counterpart.

There is a very simple model leading to a Lagrangian of the form (3.42). Consider, in the classical setup for the Aharonov-Bohm effect [3,56,76,87], the interaction of a charged particle with a flux tube . "Modulo" gauge transformations, the vector potential is, using polar coordinates (r,ϕ) in the plane:

$$\vec{A}(r,\phi) = \frac{\Phi}{2\pi r}\, \hat{\phi}\ \equiv \frac{\Phi}{2\pi}\, \nabla\phi \tag{3.45}$$

where Φ is the total flux associated with the flux tube, and $\hat{\phi}$ is a unit vector in the azimuthal direction. If we denote by e^{*} the charge of the particle, the Lagrangian will be:

$$\mathcal{L} = \frac{m\vec{v}^2}{2} - \frac{e^*}{c}\vec{A}\cdot\vec{v} \equiv \mathcal{L}_0 - \frac{e^*\Phi}{2\pi c}\frac{d\phi}{dt} ; \quad \mathcal{L}_0 = \frac{m\vec{v}^2}{2} \qquad (3.46)$$

Consider now two identical composites, made each of a particle of charge e^* **and** a flux tube with flux Φ, We are considering therefore flux tubes **rigidly** attached to each particle. Setting $c=1$, we can write the interaction Lagrangian between two such composites as:

$$\mathcal{L}_{int} = -\frac{e^*\Phi}{\pi}\frac{d\phi}{dt} \qquad (3.47)$$

where ϕ is now the **relative** azimuthal angle. Note that **(3.47) differs by a factor of 2 from the corresponding interaction term in (3.46).** This is due to the fact that in (3.47) we have taken into account the fact that **two** charges interact with **two** flux tubes, and this leads to the doubling of the interaction Lagrangian. Actually, in order to avoid complications arising from electrostatic interactions, we should consider the limiting case in which $e^* \to 0$ and $\Phi \to \infty$ in such a way that the product $e^*\Phi$ remains constant.

Setting from now on $\hbar=1$, comparison of (3.47) and (3.42) shows that (generic) *anyons can be considered as particles carrying a fictitious charge e^* and a fictitious flux tube with flux Φ, mutually interacting via*

Aharonov-Bohm-type couplings, and such that the fictitious charge and flux are related to the statistical angle θ by:

$$e^* \Phi = \theta \tag{3.48}$$

This is the first dynamical model that has been proposed in the literature to interpret the unusual properties of anyons [6]. In the next Chapter I will discuss more sophisticated models that allow to implement fractional statistics in the form (3.42).

Remark: If we reinstate $\hbar$ and c in their proper places, Eq. (3.48) becomes:

$$\frac{\Phi}{\Phi_0^*} = \frac{\theta}{2\pi} \tag{3.49}$$

where:

$$\Phi_0^* =: \frac{hc}{e^*} \tag{3.50}$$

is the **flux quantum,** or the **elementary fluxon,** associated with the charge e^*.

Therefore:

$$\frac{\theta}{2\pi} = (\# \text{ of flux quanta associated with the particles}) \qquad (3.51)$$

In this picture, Bosons will have an **integer** number and Fermions a **half-odd-integer** number of flux quanta associated with them.

In the following, we will stick however to $\hbar=c=1$, so that the flux quantum will be written as:

$$\Phi_0{}^* = \frac{2\pi}{e^*} \quad (\hbar=c=1) \qquad (3.52)$$

CH.4. TOPICS IN CHERN-SIMONS PHYSICS.

4.1. INTRODUCTION.

We have seen that a system of anyons (particles obeying fractional statistics) with statistical parameter θ can be described by modifying the (classical) particle Lagrangian $\mathcal{L}$ as:

$$\mathcal{L} \rightarrow \mathcal{L} - \frac{\theta}{\pi} \frac{d}{dt} \sum_{\alpha < \beta} \phi_{\alpha\beta}(t) \tag{4.1}$$

and that -θ- can be "read" in terms of a fictitious charge -q- and a fictitious flux Φ associated with each particle, such that

$$\theta = q\Phi \tag{4.2}$$

We shall discuss now a simple field-theoretic model [15,17,33,48,61-64,106,107,115] which will allow us to implement fractional-statistics terms of the form (4.1) via interaction of the particles with an abelian gauge field (we will not belabor here on the non-abelian generalizations, which are of lesser interest for problems in Condensed-Matter theory). What we will discuss in the following is peculiar to (2+1)-dimensional theories.

4.2. CHERN-SIMONS LAGRANGIANS.

The **Chern-Simons (CS)** [29] term in the Lagrangian (density) for a $U(1)$ gauge theory is:

$$\mathcal{L}_{CS}= \frac{k}{4\pi}\, \epsilon^{\mu\nu\lambda}A_\mu\partial_\nu A_\lambda \tag{4.3}$$

(**Note:** We employ here (Lorentz) indices $\mu=(0,1,2)$ with $\mu=0$ the time index, and the metric tensor: $g_{\mu\nu}=$ diag(1,-1,-1)). Introducing the (tensor) field strength:

$$F_{\mu\nu}= \partial_\mu A_\nu- \partial_\nu A_\mu \tag{4.4}$$

we can also write $\mathcal{L}_{CS}$ as:

$$\mathcal{L}_{CS}= \frac{k}{8\pi}\, \epsilon^{\mu\nu\lambda}A_\mu F_{\nu\lambda} \tag{4.5}$$

$\mathcal{L}_{CS}$ can be uniquely associated to the three-form:

$$\omega_{CS}= \frac{k}{4\pi}\, (\epsilon^{\mu\nu\lambda}A_\mu\partial_\nu A_\lambda)\Omega \tag{4.6}$$

where Ω is the standard volume-form:

$$\Omega= dx^0 \wedge dx^1 \wedge dx^2 \tag{4.7}$$

Equivalently:

$$\omega_{CS} = \frac{k}{4\pi} \, A \wedge dA \tag{4.8}$$

where $dA = F = \frac{1}{2} F_{\mu\nu} dx^{\mu} \wedge dx^{\nu}$ is the two-form associated with the field strength(s).

Under a gauge transformation:

$$A \rightarrow A + d\Phi \tag{4.9}$$

$(\Phi \in \mathcal{F}(\mathbb{R}^3))$, we have:

$$\omega_{CS} \rightarrow \omega_{CS} + \frac{k}{4\pi} \, d(\Phi dA) \tag{4.10}$$

In coordinates:

$$d(\Phi dA) = \epsilon^{\mu\nu\lambda}(\partial_{\mu}\Phi)\partial_{\nu}A_{\lambda}\Omega \equiv \partial_{\mu}(\epsilon^{\mu\nu\lambda}\Phi\partial_{\nu}A_{\lambda})\Omega \tag{4.11}$$

In other words, $\mathcal{L}_{CS}$ changes by a *total divergence*. The (classical) equations of motion for a (any) Lagrangian containing a CS term are therefore *gauge-invariant* (as far as the CS term is concerned).

Let's consider now the effect of adding a CS term to the standard $\mathsf{U}(1)$ Lagrangian, i.e. let's consider the Lagrangian density:

$$\mathcal{L} = -\frac{1}{4} F_{\mu\nu}F^{\mu\nu} + \mathcal{L}_{CS} \tag{4.12}$$

As [40]:

$$\frac{\partial}{\partial(\partial_\mu A_\nu)} F_{\alpha\beta}F^{\alpha\beta} = 4F^{\mu\nu} \tag{4.13}$$

and

$$\frac{\partial\mathcal{L}_{CS}}{\partial(\partial_\mu A_\nu)} = \frac{k}{4\pi}\, \epsilon^{\mu\nu\lambda}A_\lambda; \quad \frac{\partial\mathcal{L}_{CS}}{\partial A_\nu} = \frac{k}{4\pi}\, \epsilon^{\nu\lambda\rho}\partial_\lambda A_\rho \tag{4.14}$$

the standard equations of motion

$$\partial_\mu \frac{\partial\mathcal{L}}{\partial(\partial_\mu A_\nu)} - \frac{\partial\mathcal{L}}{\partial A_\nu} = 0 \tag{4.15}$$

yield

$$\partial_\mu F^{\mu\nu} = -\frac{k}{4\pi}\, \epsilon^{\nu\lambda\rho}\, F_{\lambda\rho} \tag{4.16}$$

The field equations (4.16) are gauge-invariant, as already stated. The Bianchi identity [30]:

$$\epsilon^{\nu\lambda\rho}\partial_\nu F_{\lambda\rho} = 0 \tag{4.17}$$

follows (also) from taking the divergence of both sides of Eq. (4.16).

Remark: Decreasing by one the space dimensions brings about peculiar features one should not forget about. Formally, they mostly come from the fact that, as compared with (3+1)-dimensional theories, the Ricci tensor has one index less. This implies [64] that the curl of a vector field $\vec{V}$, i.e.:

$$\nabla \times \vec{V} =: \epsilon^{ij} \partial_i V_j \qquad (4.18)$$

is indeed a ***scalar***, while one can define the "curl of a scalar" S as:

$$(\nabla \times S)^i = \epsilon^{ij} \partial_j S \qquad (4.19)$$

These results are quite obvious when stated in the language of Differential Geometry [30]. S being a scalar, dS is a one-form. As the "Lorentz" metric induces an Euclidean metric $\tilde{g}$ on the 2D underlying space, to any vector field $\vec{V}$ we can associate in a unique way a one-form:

$$\omega_{\vec{V}} =: \tilde{g}(\vec{V}, \cdot) \qquad (4.20)$$

Then, $d\omega_{\vec{V}}$ is a two-form, and (4.18-19) are nothing but a manifestation of Hodge's duality [30] as applied to $d\omega_{\vec{V}}$ and dS respectively.

One of the most striking consequences of this reduction of dimension is that, while in a (3+1)-dimensional spacetime the Bianchi identity generates the full set of the four homogeneous Maxwell equations, here (4.17) is ***a single scalar equation***. When spelled out explicitly, it reads:

$$\partial_0 F_{12} - \partial_2 F_{10} + \partial_1 F_{20} = 0 \qquad (4.21)$$

This is the 2D scalar analog of the 3D (vector) Maxwell equation:

$$\frac{1}{c} \frac{\partial}{\partial t} \vec{B} + \nabla \times \vec{E} = 0 \qquad (4.22)$$

There are no equations left, and the most interesting consequnce of this is that **the "fourth" Maxwell equation** (i.e.: $\nabla \cdot \vec{B} = 0$) **is altogether absent from the theory.**

This is a peculiarity of "2D (classical) electrodynamics", and should not be forgotten. In particular, later on we will find that the effect of the coupling of point particles with a CS field is to attach, so-to-speak, flux tubes to particles. In view of what we have just said, problems such as that of the "return flux" of such tubes, which have plagued for years the theoretical discussions of the Aharonov-Bohm effect, are entirely out of place and devoid of any content in a (2+1)-dimensional context.

In two space dimensions, the field tensor dual to $F_{\mu\nu}$ is actually a Lorentz vector:

$$^*F^\nu = \frac{1}{2} \epsilon^{\nu\lambda\rho} F_{\lambda\rho} \qquad (4.23)$$

and

$$F^{\mu\nu} = \epsilon^{\mu\nu\sigma} \, {}^*F_\sigma \qquad (4.24)$$

while the Bianchi identity becomes

$$\partial_\nu \, {}^*F^\nu = 0 \qquad (4.25)$$

Eq. (4.16) can be rewritten as:

$$\partial_\mu \epsilon^{\mu\nu\sigma} \, {}^*F_\sigma = -\frac{k}{2\pi} \, {}^*F^\nu \qquad (4.26)$$

Multiplying both sides by $\epsilon_{\alpha\beta\nu}$ and summing over ν, we obtain:

$$\partial_\alpha \, {}^*F_\beta - \partial_\beta \, {}^*F_\alpha = \frac{k}{2\pi} \epsilon_{\alpha\beta\nu} \, {}^*F^\nu \qquad (4.27)$$

or, equivalently:

$$\partial_\alpha \, {}^*F_\beta - \partial_\beta \, {}^*F_\alpha = \frac{k}{2\pi} F_{\alpha\beta} \qquad (4.28)$$

Applying ∂^α, and using the Bianchi identity (4.25), we obtain:

$$(\partial^\alpha \partial_\alpha) \, {}^*F_\beta = -\frac{k}{2\pi} \partial^\alpha F_{\alpha\beta} \qquad (4.29)$$

Then, using again the equations of motion, we find eventually:

$$(\partial^\alpha \partial_\alpha) \, {}^*F_\beta = -\frac{k^2}{4\pi^2} \, {}^*F_\beta \qquad (4.30)$$

It appears therefore that, under the effect of a CS term, *the U(1) gauge field acquires a mass term.* This is similar (though not quite the same thing) to what happens in the theory of superconductivity (see Ch.3), where the Higgs phenomenon renders the photon massive and leads to the Meissner effect.

Returning now to Eq. (4.26), the r.h.s. can be interpreted formally as a source term for the field equations. Defining then formally a "current":

$$j^\nu = -\frac{k}{2\pi}{}^*F^\nu \qquad (4.31)$$

Eq. (4.28) can be rewritten as:

$$\partial_\alpha j_\beta - \partial_\beta j_\alpha = -\frac{k^2}{4\pi^2}F_{\alpha\beta} \qquad (4.32)$$

In the theory of superconductivity, the London equation (cfr. Eq. (2.69)):

$$\nabla \times \vec{j} = -\frac{c}{4\pi\lambda^2}\vec{B} \qquad (4.32')$$

relates the current $\vec{j}$ to the magnetic field strength $\vec{B}$, λ being the London penetration depth (see Sect. 2.5).

We have been working here in natural units in which c=1 and, moreover (see Eq. (4.31)), we are writing the field equations without the customary factor of 4π in the r.h.s. (rationalized units). With these minor modifications in mind, Eq. (4.32) appears as a (formal) generalization of the London equation (4.32'), and the penetration depth is now given by:

$$\lambda^2 = \frac{4\pi^2}{k^2} \qquad (4.33)$$

4.3. CHARGED PARTICLES INTERACTING WITH A CS FIELD.

Consider [16,17], in two space dimensions, a set of identical particles with charge -e- and mass -m-, interacting, via minimal coupling, with a $\mathbb{U}(1)$ gauge field. The "charge" of the particles has here only the meaning of a coupling constant for the interaction with the gauge field, and has nothing to do with physical electrical charge. We will label the particles by an index: $\alpha=1...N$ (not to be confused with a Lorentz index!) and the particle positions by a vector $\vec{r}_\alpha \in \mathbb{R}^2, \alpha=1...N$. We will also sometimes employ the "Lorentz" notation:

$$z_\alpha{}^\mu =: (z_\alpha{}^0, z_\alpha{}^1, z_\alpha{}^2) \equiv (t, \vec{r}_\alpha) \qquad (4.34)$$

Note however that, despite this notation which will turn out at times to be convenient, the particles will be assumed for simplicity to be nonrelativistic. The particle Lagrangian is then:

$$L = \frac{m}{2} \sum_\alpha [\frac{d\vec{r}_\alpha}{dt}]^2 - e \sum_\alpha \dot{z}_\alpha{}^\mu A_\mu(z_\alpha) + \int \mathcal{L}_{field} d^2x \qquad (4.35)$$

For the field, we take:

$$\mathcal{L}_{field} \equiv \mathcal{L}_{CS} \tag{4.36}$$

The neglect of terms higher in derivatives in the field Lagrangian has been motivated in various ways in the literature [17,106].

We can introduce the particle four-current as:

$$j^{\mu} =: \sum_{\alpha} \dot{z}_{\alpha}^{\mu} \, \delta^{(2)}(\vec{x} - \vec{r}_{\alpha}(t)) \tag{4.37}$$

In this way, the Lagrangian can be rewritten as:

$$L = \frac{m}{2} \sum_{\alpha} [\frac{d\vec{r}_{\alpha}}{dt}]^2 + \int d^2x \mathcal{L}_A \tag{4.38}$$

where:

$$\mathcal{L}_A = -ej^{\mu}A_{\mu} + \mathcal{L}_{CS} \tag{4.39}$$

We will use the (cumbersome) specific form (4.37) of j^{μ} only when strictly necessary. As far as the field dynamics is concerned, the only requisite of j^{μ} that we will need is that it be a **conserved** current, i.e. that:

$$\partial_{\mu}j^{\mu} = 0 \tag{4.40}$$

The particles' equations of motion can be derived in a standard way, and they read:

$$m\ddot{z}_{\alpha,i} = eF_{i\lambda}(z)\dot{z}_{\alpha}^{\lambda} \tag{4.41}$$

(as customary, we employ Latin indices for space coordinates (i.e.: i=1,2), Greek ones for space-time coordinates, ranging from 0 to 2). From the field Lagrangian $\mathcal{L}_A$ we obtain:

$$\frac{\partial \mathcal{L}_A}{\partial(\partial_\mu A_\nu)} \equiv \frac{\partial \mathcal{L}_{CS}}{\partial(\partial_\mu A_\nu)} \; ; \; \frac{\partial \mathcal{L}_A}{\partial A_\nu} = \frac{\partial \mathcal{L}_{CS}}{\partial A_\nu} - ej^\nu \qquad (4.42)$$

and hence, with some elementary algebra, the field equations:

$$\frac{k}{4\pi} \epsilon^{\mu\nu\lambda} F_{\nu\lambda} = ej^\mu \qquad (4.43)$$

or, equivalently:

$$F_{\mu\nu} = \frac{2\pi e}{k} \epsilon_{\mu\nu\rho} j^\rho \qquad (4.44)$$

Substituting (4.44) back into (4.41), and using (4.37), we find at once:

$$\ddot{\vec{r}}_\alpha = 0 \qquad (4.45)$$

i.e., *at the classical level*, the particles are *free. Classically, the gauge field does not generate any kind of forces.* Let us stress, at the risk of being pedantic, that this is true purely at the *classical* level. Quantum-mechanically, it will be seen tha the CS term generates topological interactions of the Aharonov-Bohm type among the particles, and hence fractional statistics.

Remark: Eq. (4.43) for $\mu=0$ (or (4.44) for $\mu=1$, $\nu=2$) does not contain time derivatives. It is therefore not an evolution equation, but rather a **constraint equation** (Gauss law constraint):

$$ej^0 = \frac{k}{2\pi} F_{12} \tag{4.46}$$

F_{12} is the "magnetic field" B associated with the gauge field. Integrating over the (x^1, x^2) plane, we find:

$$eN = \frac{k}{2\pi} \Phi \tag{4.47}$$

where Φ is the total flux associated with B. *Therefore, to each particle there is associated a flux:*

$$\phi = \frac{2\pi e}{k} \tag{4.48}$$

What Eq. (4.43) tells us is that (with the approximation of retaining only $\mathcal{L}_{CS}$) *the gauge field has no independent dynamics.* The field dynamics is entirely determined by that of the particles via the current on the r.h.s. of (4.43). Moreover, if j_μ is of the form (4.37), we see that *the field strength* (i.e. the gauge-invariant content of the gauge field) *is entirely concentrated at the location of the particles* (or, in any event, that $F_{\mu\nu}$ is entirely concentrated on the support of the four-current).

The picture that emerges is that *the main effect of the CS term is to attach to each particle a flux tube (the flux being given by (4.48)), and that such flux tubes are carried around by the motion of the particles in a "rigid" way.*

Let us now substitute back the field strength as given by (4.43) into the form (4.5) of the CS Lagrangian. We find at once:

$$\mathcal{L}_{CS} = \frac{k}{8\pi} \epsilon^{\mu\nu\lambda} A_\mu F_{\nu\lambda} \equiv \frac{e}{2} j^\mu A_\mu \qquad (4.49)$$

Hence:

$$\mathcal{L}_A = -\frac{e}{2} j^\mu A_\mu \qquad (4.50)$$

and the net effect of the Chern-Simons term is to cancel **half** of the minimal-coupling interaction (and hence also **half** of the flux associated with each particle).

Remark: That the equations of motion for the field actually generate a constraint can be better seen by abandoning the "manifestly covariant" notation employed up to now. We find then:

$$\mathcal{L}_A = -eA_0 j_0 + e\vec{A} \cdot \vec{j} + \mathcal{L}_{CS} \qquad (4.51)$$

The CS Lagrangian can be rearranged as follows:

$$L_{CS} = \int d^2x \mathcal{L}_{CS} = \frac{k}{4\pi} \int d^2x \left\{ A_0 \epsilon^{ij} \partial_i A_j + A_1 \partial_2 A_0 - A_2 \partial_1 A_0 + \right.$$

$$\left. + \epsilon^{ij} A_i \partial_0 A^j \right\} \qquad (4.52)$$

Furthermore:

$$\int d^2x (A_1 \partial_2 A_0 - A_2 \partial_1 A_0) = \int d^2x (\partial_2(A_1 A_0) - \partial_1(A_2 A_0) +$$

$$+ A_0 \epsilon^{ij} \partial_i A_j) \quad (4.53)$$

Therefore, apart from (vanishing) boundary terms:

$$L_{CS} = \frac{k}{4\pi} \int d^2x \, \epsilon^{ij} A_i \partial_0 A_j + \frac{k}{2\pi} \int d^2x \, A_0 \epsilon^{ij} \partial_i A_j \quad (4.54)$$

and an equivalent field Lagrangian (density) will be:

$$\tilde{\mathcal{L}}_A = e\vec{A} \cdot \vec{j} + A_0(\frac{k}{2\pi} \epsilon^{ij} \partial_i A_j - ej^0) + \frac{k}{4\pi} \int d^2x \, \epsilon^{ij} A_i \partial_0 A_j \quad (4.55)$$

In this way we have made the dependence on A_0 explicit and it is quite obvious that varying w.r.t. A_0 will yield the constraint (4.46), as $\epsilon^{ij} \partial_i A_j \equiv F_{12}$. It is also interesting to remark that the Lagrangian (4.55) is at most **linear** in the time derivatives of the field components. If treated as a field Lagrangian "per se", is is therefore a constrained Lagrangian, and should be treated with the methods of Dirac's theory of constraints [14]. We will not however belabor on this point any further.

To end this subsection, let us stress that, in view of the result (4.49), if we can solve Eqns. (4.43) for A_μ, the effective Lagrangian describing the (statistical) interaction among the particles will be eventually :

$$L_{eff} = \frac{m}{2} \sum_\alpha [\frac{d\vec{r}_\alpha}{dt}]^2 - \frac{e}{2} \int d^2x \, j^\mu A_\mu \qquad (4.56)$$

or, whenever (4.37) holds:

$$L_{eff} = \frac{m}{2} \sum_\alpha [\frac{d\vec{r}_\alpha}{dt}]^2 - \frac{e}{2} \sum_\alpha \dot{z}_\alpha^{\ \mu} A_\mu(z_\alpha) \qquad (4.57)$$

4.4. GAUGE FIXING AND AN EXPLICIT SOLUTION FOR A_μ.

Let's go back now to the field equations, namely:

$$\frac{k}{2\pi} F_{\mu\nu} = e\epsilon_{\mu\nu\lambda} j^\lambda \qquad (4.58)$$

Consider first the constraint equation:

$$\frac{k}{2\pi} B = ej^0 \; ; \quad B = F_{12} \equiv \partial_1 A_2 - \partial_2 A_1 \qquad (4.59)$$

and the simple case in which j^0 is entirely concentrated at the origin:

$$j^0 = \delta^{(2)}(\vec{x}) \qquad (4.60)$$

This corresponds to an Aharonov-Bohm solenoid (actually a single flux line) at the origin, with flux $2\pi e/k$. We know that, "modulo" gauge transformations, the associated vector potential is:

$$\vec{A} = \frac{\phi}{2\pi} \nabla\theta \; ; \quad \phi = \frac{2\pi e}{k} \tag{4.61}$$

where θ is the azimuthal angle of the potential point in a polar coordinate system centered at the origin. In terms of Cartesian coordinates:

$$\vec{A} = \frac{\phi}{2\pi} \frac{\hat{z} \times \vec{r}}{r^2} \Leftrightarrow A_i(\vec{x}) = \frac{\phi}{2\pi} \epsilon_{ij} \frac{x^j}{r^2} \tag{4.62}$$

where $\hat{z}$ is a unit vector perpendicular to the plane, and $r = |\vec{x}|$.

By Stokes' theorem, Eqns.(4.61-62) imply that the associated magnetic field be in the $\hat{z}$ direction, $\vec{B} = B\hat{z}$, and that:

$$B = \phi\delta^{(2)}(\vec{x}) \tag{4.63}$$

Therefore, with the identification $\phi = 2\pi e/k$, (4.61) solves (4.59) for a single source. The solution for a general j^0 is now easily found by superposition (after all, (4.63) is nothing but the Green function for the general inhomogeneous problem (4.59)), and the solution is:

$$A_i(\vec{x},t) = \frac{e}{k} \int d^2y \; \epsilon_{ij} \frac{(\vec{x}-\vec{y})^j}{|\vec{x}-\vec{y}|^2} j^0(\vec{y},t) \tag{4.64}$$

or, in vector notation:

$$\vec{A}(\vec{x},t)= \frac{e}{k} \int d^2y \; [\nabla_x \theta(\vec{x}-\vec{y})] \, j^0(\vec{y},t) \tag{4.65}$$

As one can easily show that:

$$\nabla^2_x \theta(\vec{x}-\vec{y})=0 \tag{4.66}$$

the solution (4.65) implies that we have implicitly chosen the gauge:

$$\nabla \cdot \vec{A} \equiv \partial_i A_i = 0 \tag{4.67}$$

Explicitly, with the form (4.37) for j^0:

$$A_i(\vec{x},t)= \frac{e}{k} \sum_\alpha \epsilon_{ij} \frac{(\vec{x}-\vec{r}_\alpha(t))^j}{|\vec{x}-\vec{r}_\alpha(t)|^2} \tag{4.68}$$

or, in a more compact form:

$$\vec{A}(\vec{x},t)= \frac{e}{k} \sum_\alpha \vec{\nabla}_x \theta(\vec{x}-\vec{r}_\alpha(t)) \tag{4.69}$$

where $\theta(\vec{x}-\vec{r}_\alpha)$ is the **relative** azimuthal angle between $\vec{x}$ and $\vec{r}_\alpha$.

Remark: Just as the elementary solution (4.61) becomes singular for $\vec{x} \to$ 0, so does (4.68) (or (4.69) for that matter) whenever $\vec{x} \to \vec{r}_\alpha$ for some $\alpha = 1...N$. As there are no singularities of Eq. (4.65) as long as $j^0(\vec{y})$ is well-behaved, *this is clearly an artifact of having approximated the distribution of matter with a distribution of point-particles.* As is customary, we shall "cure" this pathology *by excluding* (statistical) *self-interactions.* That is, we shall write, whenever we will need $\vec{A}$ at any of the particles' positions:

$$\vec{A}(\vec{r}_\alpha) = \frac{e}{k} \sum_{\beta \neq \alpha} \nabla_\alpha \theta(\vec{r}_\alpha - \vec{r}_\beta) \tag{4.70}$$

Also (cfr. Ch. 3), when dealing with identical particles, we exclude the possibility that $\vec{r}_\beta = \vec{r}_\alpha$ for some $\beta \neq \alpha$, and hence further possible sources of ambiguity.

For general $\vec{x} \neq \vec{r}_\alpha$ $\forall \alpha$, (4.69) can be properly viewed as the gauge potential felt by an **extra** text particle.

The elementary solution (4.61) satisfies the Coulomb gauge condition (4.67), and so does the general solution we have just found for an arbitrary charge distribution. Therefore, (4.65) (or (4.67)) represents the solution for the *space* part of A_μ in the Coulomb gauge. By the very way by which it has been obtained, it solves by construction the Gauss-law constraint equation (4.59). When written explicitly, the remaining equations from the set (4.58) are:

$$F_{01} \equiv \partial_0 A_1 - \partial_1 A_0 = \frac{2\pi e}{k} j^2 \tag{4.71-a}$$

and:

$$F_{20} \equiv \partial_2 A_0 - \partial_0 A_2 = \frac{2\pi e}{k} j^1 \tag{4.71-b}$$

As $\partial_0 A_i$ (i=1,2) is known from the solution that we have just found, we are led to the following pair of equations for A_0:

$$\partial_1 A_0 = \partial_0 A_1 - \frac{2\pi e}{k} j^2 \tag{4.76-a}$$

and

$$\partial_2 A_0 = \partial_0 A_2 + \frac{2\pi e}{k} j^1 \tag{4.76-b}$$

or, more compactly:

$$\nabla A_0 = \partial_0 \vec{A} + \frac{2\pi e}{k} \hat{z} \times \vec{j} \tag{4.77}$$

The consistency condition that $\partial_1 \partial_2 A_0$ be the same when derived from either one of Eqns. (4.76) leads to:

$$0 = \partial_1(\partial_0 A_2 + \frac{2\pi e}{k} j^1) - \partial_2(\partial_0 A_1 - \frac{2\pi e}{k} j^2) = \partial_0(\partial_1 A_2 - \partial_2 A_1) + \frac{2\pi e}{k}(\partial_1 j^1 + \partial_2 j^2) =$$

$$= \partial_0 F_{12} + \frac{2\pi e}{k}(\partial_1 j^1 + \partial_2 j^2) \tag{4.78}$$

On the other hand, the constraint equation (4.59) ($B \equiv F_{12}$) yields:

$$\partial_0 F_{12} = \frac{2\pi e}{k} \partial_0 j^0 \qquad (4.79)$$

and the consistency condition becomes:

$$0 = \frac{2\pi e}{k} \partial_\mu j^\mu \qquad (4.80)$$

Therefore, *the equations for A_0 are mutually consistent iff j^μ is a conserved current.* That the CS field be coupled to a *conserved* current appears therefore as an *internal consistency requirement for the whole theory.*

After these preliminaries, we show now that [16,17], "modulo" gauge transformations (and actually within the gauge (4.67)), *the general solution of the field equations (4.58) is given by:*

$$A_\mu(\vec{x},t) = \frac{e}{k} \partial_\mu \int d^2y \; \theta(\vec{x}-\vec{y}) j^0(\vec{y},t) \qquad (4.81)$$

where $\theta(\vec{x}-\vec{y})$ is the relative azimuthal angle between the points $\vec{x}$ and $\vec{y}$ $\in \mathbf{R}^2$.

Remark: An expression like: "$A_\mu = \partial_\mu$(something)" seems to imply that A_μ is a pure gauge and that it can be done away with by means of a (regular) gauge transformation. That this is not actually so stems from the fact that the "something" in (4.81) is an angle, which is not in itself a globally defined function, while its gradient is. A_μ can then be gauged away only by means of a *singular* gauge transformation of the kind discussed in Ch.3.

That (4.81) solves indeed (4.58) follows from the fact that j^μ is a conserved current, and from the identity (to be understood in the distribution sense):

$$(\partial_i\partial_j-\partial_j\partial_i)\theta(\vec{x}-\vec{y})= 2\pi\,\delta^{(2)}(\vec{x}-\vec{y}) \qquad (4.82)$$

($\partial_i \equiv \partial/\partial x^i$ etc.). A quick way to prove (4.82) is to observe that:

$$\partial_i\theta(\vec{x}-\vec{y}) = -\epsilon_{ij}\partial_j\ln|\vec{x}-\vec{y}| \qquad (4.83)$$

which leads to:

$$(\partial_i\partial_j-\partial_j\partial_i)\theta(\vec{x}-\vec{y})= \epsilon_{ij}\nabla^2\ln|\vec{x}-\vec{y}| \qquad (4.84)$$

But it is well known that:

$$\nabla^2\ln|\vec{x}-\vec{y}|= 2\pi\,\delta^{(2)}(\vec{x}-\vec{y}) \qquad (4.85)$$

and this proves (4.82).

Then, we prove that:

$$(\partial_\mu \partial_\nu - \partial_\nu \partial_\mu) \int d\vec{y} \ \theta(\vec{x}-\vec{y}) j^0(y) = 2\pi \ \epsilon_{\mu\nu\rho} j^\rho(x) \qquad (4.86)$$

(here: $x \equiv (\vec{x},t)$, $y \equiv (\vec{y},t)$). For $\mu,\nu = i,j$, (4.86) is a straightforward consequence of (4.82). We observe next that, by using the conservation of the current and neglecting boundary terms arising from a partial integration, one proves easily that:

$$\partial_0 \int d\vec{y} \ \theta(\vec{x}-\vec{y}) j^0(y) = -\partial_k \int d\vec{y} \ \theta(\vec{x}-\vec{y}) \ j^k(y) \qquad (4.87)$$

So, finally, (4.87) and (4.82) allow us to prove (4.86) also in the case in which one of the indices is a time index, and this achieves the proof. $\square$

Having solved for the gauge field in terms of the matter field, we will proceed, in the next Section, to derive the effective Lagrangian describing the statistical interparticle interaction induced by the Chern-Simons term.

4.5. EFFECTIVE LAGRANGIAN FOR (STATISTICAL) PARTICLE INTERACTION.

We can combine now the results of the previous Sections in order to obtain an explicit description of the statistical interparticle interaction.

As discussed in Sect. 4.3 (cfr. Eq. (4.56)), the effective Lagrangian is:

$$L_{eff} = \frac{m}{2} \sum \left[\frac{d\vec{r}_\alpha}{dt}\right]^2 - \frac{e}{2} \int d^2x \, j^\mu(\vec{x},t) A_\mu(\vec{x},t) \qquad (4.88)$$

Using the results of Sect. 4.4, we obtain:

$$L_{int} = -\frac{e}{2} \int d\vec{x} \, j^\mu(x) \, A_\mu(x) \equiv -\frac{e^2}{2k} \int d\vec{x} d\vec{x}' \, j^\mu(x) \partial_\mu \theta(\vec{x}-\vec{x}') \, j^0(x') \qquad (4.89)$$

$(x \equiv (\vec{x},t), \, x' \equiv (\vec{x}',t))$ or, more explicitly:

$$L_{int} = -\frac{e^2}{2k} \int d\vec{x} d\vec{x}' \left\{ j^0(x) \, \theta(\vec{x}-\vec{x}') \, \partial_0 j^0(x') + \vec{j}(x) \cdot \nabla_x \theta(\vec{x}-\vec{x}') \, j^0(x') \right\} \qquad (4.90)$$

Integrating by part the second term, neglecting boundary terms and using again the continuity equation, we get:

$$L_{int}= -\frac{e^2}{2k} \int d\vec{x}d\vec{x}'\left\{j^0(x)\theta(\vec{x}-\vec{x})\partial_0 j^0(x')+ \partial_0 j^0(x)\theta(\vec{x}-\vec{x}')\, j^0(x')\right\}$$

$$(4.91)$$

All in all, this amounts to writing the interaction Lagrangian as:

$$L_{int}=- \frac{e^2}{2k} \frac{d}{dt} \int d\vec{x}d\vec{x}\, j^0(x)\theta(\vec{x}-\vec{x}')j^0(x') \qquad (4.92)$$

i.e. explicitly as a **total time derivative**. This emphasizes once again that the particles are classically free, but, as we already know, total time derivatives of **angles** lead to topological effects and to fractional statistics in the quantum version of the problem.

At this stage, we can safely use Eq. (4.32). Again neglecting the self-interaction terms, we obtain:

$$L_{int}=- \frac{e^2}{2k} \sum_{\alpha \neq \beta} \frac{d}{dt} \theta_{\alpha\beta} \; ; \quad \theta_{\alpha\beta}= \theta(\vec{r}_\alpha-\vec{r}_\beta) \qquad (4.93)$$

Equivalently (as it is easy to show that: $\theta_{\alpha\beta}=\theta_{\beta\alpha}$ mod.π):

$$L_{int}=- \frac{e^2}{k} \sum_{\alpha < \beta} \frac{d}{dt}\theta_{\alpha\beta} \qquad (4.94)$$

So, we end up with the effective Lagrangian:

$$L_{eff} = \frac{m}{2} \sum_{\alpha} \left[\frac{d\vec{r}_\alpha}{dt}\right]^2 - \frac{e^2}{k} \sum_{\alpha < \beta} \frac{d}{dt} \theta_{\alpha\beta} \qquad (4.95)$$

Comparison with (4.1) shows then that *the effective Lagrangian (4.101) describes a system of anyons characterized by the statistical angle:*

$$\theta = \frac{\pi e^2}{k} \qquad (4.96)$$

A Hamiltonian description of our system can be obtained easily from the effective Lagrangian (4.95). The conjugate momenta become:

$$\vec{p}_\alpha =: \frac{\partial L_{eff}}{\partial \dot{\vec{r}}_\alpha} = m\dot{\vec{r}}_\alpha - \frac{e^2}{k} \sum_{\beta}{}' \nabla_\alpha \theta_{\alpha\beta} \qquad (4.97)$$

where the $\sum'$ stands for a sum over all the particle indices $\beta \neq \alpha$. The Hamiltonian is then :

$$\mathcal{H} = \frac{1}{2m} \sum_{\alpha} \left\{\vec{p}_\alpha - e\vec{a}(\vec{r}_\alpha)\right\}^2 \qquad (4.98)$$

where:

$$\vec{a}(\vec{r}_{\alpha})=: -\frac{e}{k} \sum_{\beta}{}' \nabla_{\alpha} \theta_{\alpha\beta} \qquad (4.99)$$

Considering (4.98-99) a bit more closely, it looks as if we had been performing some kind of black magic by which the Hamiltonian contains only a vector potential in the standard minimal-coupling combination with the canonical momentum, while the scalar potential has been done away with and is altogether absent. Nonetheless, (4.98) , having been derived by elementary and standard means, is obviously an **exact** result. The solution of the puzzle is however rather simple. Indeed, it can be proved with some algebraic effort that:

$$\int d\vec{x}\, j^0(x) A_0(x) \equiv - \int d\vec{x}\, \vec{j}(x) \cdot \vec{A}(x) \ . \qquad (4.100)$$

with A_0 and $\vec{A}$ given by (4.81). More explicitly, using again the fact that j^{μ} is a conserved current and neglecting boundary terms arising from partial integrations:

$$\int d\vec{x} d\vec{x}'\, j^0(x)\theta(\vec{x}-\vec{x}')\partial_0 j^0(x') \equiv \int d\vec{x} d\vec{x}'\, \vec{j}(x) \cdot \nabla_x \theta(\vec{x}-\vec{x}')\, j^0(x') \equiv$$

$$\equiv \frac{1}{2}\frac{d}{dt}\int d\vec{x} d\vec{x}'\, j^0(x)\, \theta(\vec{x}-\vec{x}')\, j^0(x') \qquad (4.101)$$

Therefore, *the scalar and vector part of the electromagnetic potential contribute each exactly <u>half</u> of the final interaction Lagrangian* (4.92).

It turns out then that we can describe the whole theory of a matter field minimally coupled to a Chern-Simons gauge field *by employing only a vector potential* of the form (4.99), one however which has *twice* the nominal value it would have in a naïve approach. This solves the apparent problems that a superficial look at (4.98) seemed to have created.

The Hamiltonian (4.98) has been the starting point [106,107] for the treatment of Quantum Mechanics of the anyon gas. In particular, the broad subject of anyon superconductivity at special values ($\theta = \pi(1-1/n)$ for integer n) of the statistical angle (4.92) has (4.98) as its starting point. These topics of current research are however outside the scope of these Lecture Notes.

We conclude here our discussion of Chern-Simons terms, contenting ourselves of having shown how fractional statistics can be rigorously implemented via a local field theory construction.

CH. 5. A SHORT INTRODUCTION TO: CONNECTIONS ON U(1) BUNDLES AND BERRY'S PHASE.

5.1. INTRODUCTION.

This Chapter is meant only to introduce in an elementary way some ideas associated with connections on principal fiber bundles [30,31,45,75]. Actually, I shall consider only U(1) bundles, and not general principal bundles, as are only the former that will be needed in the following. A notable U(1) bundle that we have already encountered in Chapts. 1 and 2 is the *Hopf bundle* which is associated with the Hopf fibration $S^3 \rightarrow S^2$. It turns out that it is precisely this mathematical structure that is associated with the (classical) optical phenomenon discovered by Pancharatnam [81] and that goes under the name of "Pancharatnam's phase" [20,21,22]. I will discuss next some ideas connected with the "adiabatic" or "Berry's" phase [18] that will turn out to be useful in the discussion, in the next Chapter, of some relevant aspects of the Quantum Hall Effect (QHE). In the context of Berry's phase, I will discuss in an elementary way connections on a Hilbert bundle, having in mind the Hilbert space of states of a Hamiltonian depending on some set of parameters. Technically speaking, I will deal there with an **associated** [30] bundle rather than with a principal U(1) bundle directly, but the construction will turn out to be essentially similar to that employed for the Hopf bundle. I will make reference mainly to the material, both original and reprinted, contained in [27,75,88].

5.2. LIGHT POLARIZATION, THE HOPF BUNDLE AND PANCHARATNAM's PHASE.

Let us discuss, to begin with, how one can achieve a complete characterization of a plane polarized electromagnetic wave propagating, for simplicity, in the vacuum. Taking the x^3-axis of a Cartesian frame as the direction of wave propagation, the associated electromagnetic field can be represented as:

$$E_1= a_1\cos(\tau+\delta_1); \quad E_2= a_2\cos(\tau+\delta_2); \quad E_3=0$$

$$\vec{H}=\hat{x}^3 \times \vec{E}; \quad \tau= \omega(t - \tfrac{z}{c})$$

(5.1)

where ω is the frequency of the wave and c the speed of light. Without loss of generality, we can normalize the amplitude of the wave, i.e. we will assume in the following that:

$$(a_1)^2+(a_2)^2= 1 \tag{5.2}$$

Also, again without loss of generality, we shall assume: $a_1, a_2 \geq 0$, and, in view of (5.2), we can also define an angle θ through:

$$a_1= \cos(\theta/2); \quad a_2=\sin(\theta/2); \quad 0 \leq \theta \leq \pi \tag{5.3}$$

Generically, as is well known, the locus of the fields ($\vec{E}$ and/or $\vec{H}$) will be an ellipse inscribed in the rectangle of sides $2a_1$ and

$2a_2$. whose main principal axis makes with the x-axis an angle ψ defined by (see Fig. 19-a, which is drawn for $a_1 > a_2$) :

$$\text{tg}(2\psi)= \frac{2a_1 a_2}{a_1{}^2 - a_2{}^2} \cos\delta \; ; \; \delta=\delta_2 - \delta_1 \tag{5.4}$$

The relative phase δ can be limited to the interval: $-\pi \le \delta < \pi$, and $\delta > 0$ corresponds to **right-handed**, $\delta < 0$ to **left-handed** polarized light. Note that fixing the a_i's and the relative phase δ fixes uniquely the shape and orientation of the ellipse as well as the "handedness" of the polarization. Then, the shape and orientation of the ellipse are entirely determined by the angles θ and δ. It is customary to characterize the state of polarization of a plane wave in terms of the so-called *Stokes parameters*, which can be defined as:

$$s_1= 2a_1 a_2 \cos\delta, \quad s_2= 2a_1 a_2 \sin\delta, \quad s_3= a_1{}^2 - a_2{}^2 \tag{5.5}$$

They satisfy: $s_1{}^2 + s_2{}^2 + s_3{}^2 = 1$, and one finds easily:

$$s_1= \sin\theta \, \cos\delta, \quad s_2= \sin\theta \, \sin\delta, \quad s_3= \cos\theta \tag{5.6}$$

Therefore, the states of polarization of a plane wave are represented (bijectively) by the points of a two-sphere which is known as the the *Poincare' sphere* (Fig. 19-b). In particular, the great

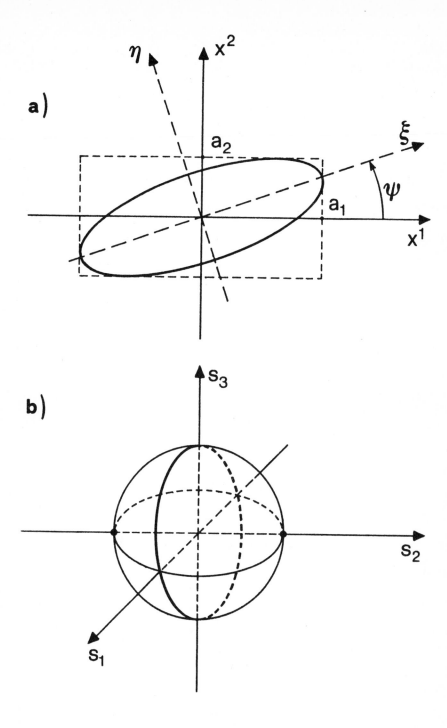

Fig. 19. a) The ellipse describing a plane polarized wave.

b) The Poincare' sphere.

circle $s_2=0$ will represents the states of **linear** polarization, while the two points $s_2=\pm 1$ will represent the two states of **circular** polarization.

The representation of the states of polarization on the Poincare' sphere lacks however one piece of information, namely the initial position of the field vectors on the ellipse describing the polarization. By reverting to a (familiar in the context of Electromagnetism) complex notation, the full information about, say, the electric field is encoded in the complex, two-component *spinor*:

$$z= \left| \begin{array}{c} z_1 \\ z_2 \end{array} \right| \tag{5.7}$$

where:

$$z_1= \cos(\theta/2)\exp[i\chi] : z_2= \sin(\theta/2)\exp[i(\chi+\delta)]; \quad \chi \equiv \delta_1 \tag{5.8}$$

Hence: $z^\dagger z=1$, and z will span the **three**-sphere S^3. The fields will be expressed as:

$$E_i= \mathrm{Re}[z_i\exp(i\tau)] \ , \ i=1,2; \tag{5.9}$$

It is quite obvious that we are recovering here the same construction of the Hopf fibration that was introduced in Sect. 2.4. In particular, the Hopf map (2.69) yields now:

$$z^\dagger \vec{\sigma}\, z = \vec{s} \qquad\qquad (5.10)$$

where the unit vector $\vec{s}$ is given by Eq. (5.6). Therefore, *in order to fully describe a plane polarized wave we need the Hopf bundle over the Poincare' sphere.*

Considering two waves, described by the spinors z and z', the (time averaged) intensity of their superposition will be given (apart from an overall factor of 2) by:

$$(z+z')^\dagger(z+z') = 2 + 2\,\mathrm{Re}(z^\dagger z') \qquad\qquad (5.11)$$

Pancharatnam [81] defined a "condition of maximum parallelism" betwen the two waves as that in which the intensity of the superposition is a maximum. Maximum parallelism will require then:

$$z^\dagger z' \geq 0 \qquad\qquad (5.12)$$

and hence, in particular, that $z^\dagger z'$ be real. We will now show how this condition can be achieved by defining a **procedure of parallel transport** on the Poincare' sphere.

Consider a path on the Poincare' sphere, and choose at will a "preimage" in S^3 of a point along the path, i.e. a spinor z projecting down to the given point under the Hopf map. The general problem of parallel transport, on our $U(1)$ bundle as well as in more general principal (and/or associated) bundles [30,31,45,75] is to define a smooth way of transporting z along a path in the bundle which projects down to the given path on the base manifold (the Poincare' sphere, in our case).

Let z and z+dz be two neighboring spinors. To lowest order in dz, normalization will require: $z^\dagger dz + dz^\dagger z = 0$, i.e. the real part of $z^\dagger dz$ will vanish by virtue of the normalization of the spinor. The parallel transport procedure will be defined now by the requirement that the imaginary part vanishes as well, i.e., all in all, by the condition:

$$z^\dagger dz = 0 \tag{5.13}$$

Comparison with Eq.(2.85) shows then that *the parallel transport procedure is being defined, in the Hopf bundle, by the condition*:

$$\mathbb{A}(\dot{z}) = 0 \; ; \; \mathbb{A} = -iz^\dagger dz \tag{5.14}$$

where $\dot{z}$ is the tangent vector to the curve along which parallel transport takes place.

Remark: That $\mathbb{A}$ is indeed a **connection one-form**, i.e. that it has the correct equivariance properties under the action of the structure group of the bundle ($\mathbb{U}(1)$ in the present case) is proved by Eq. (2.80). As we will not need here the full machinery of connections on general bundles, I refer to the literature [30,31,45,75] for a more comprehensive and exhaustive discussion.

For example, along the equator of the Poincare' sphere $(\theta=\pi/2)$, the parallel transport condition (5.14) becomes:

$$d\chi + \frac{1}{2}\, d\delta = 0 \qquad (5.15)$$

while along, say, any one of the meridians (δ=const.) of the sphere it reduces to:

$$d\chi = 0 \qquad (5.16)$$

Consider now parallel transport between two points on the Poincare' sphere along a geodesic connecting them, i.e. along one of the two arcs of the great circle passing through the two points. Without loss of generality, we may take the two points to lie on the equator of the sphere. Let then the initial point z have coordinates $(\pi/2, \delta, \chi)$, and let $(\pi/2, \delta', \chi')$ be those of the final point z'. Eq. (5.15) tells us then that:

$$\Delta\chi \equiv \chi' - \chi = -\frac{1}{2}\Delta\delta \equiv -\frac{1}{2}(\delta' - \delta) \qquad (5.17)$$

Therefore:

$$z^\dagger z' = \frac{1}{2}\, e^{i\Delta\chi}\left(1 + e^{i\Delta\delta}\right) = \cos\left(\frac{\Delta\delta}{2}\right) \qquad (5.18)$$

According to which geodesic arc we choose, $\cos(\Delta\delta/2)$ can take two opposite values. We see therefore that *Pancharatnam's maximum* *parallelism condition can be achieved by parallel transporting a plane wave between any two points on the Poincare' sphere along the <u>shortest</u> geodesic path connecting them.*

To close this Section, let us consider the effect of parallel transporting a wave along a closed path on the Poincare' sphere, and precisely along a geodesic triangle. Let A,B,C be the vertices of the triangle, and let us choose, for simplicity, A and B to lie on the equator as before, and C to be, say, the North pole. Then, circling around the triangle ABC, when returning to the initial point a plane wave has acquired a total phase of:

$$\Delta\chi = -\frac{1}{2}\Delta\delta \qquad (5.19)$$

where: $\Delta\delta = \delta(B) - \delta(A)$. But then $-\Delta\chi$ is exactly **half** of the solid angle $\Omega(ABC)$ subtended by the triangle. That decomposing a plane wave in two components, circling one of them along a geodesic triangle on the Poincare' sphere and letting then the two waves to interfere would lead to an interference effect determined by half of the solid angle subtended by the triangle is precisely the original prediction of Pancharatnam [81].

5.3. THE QUANTUM ADIABATIC PHASE.

Le M be a smooth manifold, and let the Hamiltonian $\hat{H}$ of a physical system depend on a set of parameters ranging in M:

$$\hat{H} = \hat{H}(q) \; ; \quad q \in M \tag{5.20}$$

Let $E_0 = E_0(q)$ be an eigenvalue of $\hat{H}$. *We shall assume that $E_0(q)$ is non-degenerate as q varies over M.* Actually it is not hard to show [89] that there is no loss of generality in taking $E_0 = 0$ everywhere in M. Then, we can associate with each $q \in M$ a one-dimensional Hilbert space spanned by a single, normalized eigenstate $|\phi(q)>$:

$$\hat{H}(q) | \phi(q) > \; = 0 \; ; \quad < \phi(q) | \phi(q) > \; = 1 \; \forall \, q \in M \tag{5.21}$$

As is well known, neither of Eqns. (5.21) determines the **phase** of $|\phi>$. *We shall assume that there is an open covering $\{\mathcal{U}_i\}$ of M by open sets $\mathcal{U}_i$ s.t. a smooth assignment of the phase is possible on each one of the $\mathcal{U}_i$'s.* Note that, if $\mathcal{U}_i \cap \mathcal{U}_j \neq \emptyset$, and we denote by $|\phi>_i$ and $|\phi>_j$ the basis vectors in $\mathcal{U}_i$ and $\mathcal{U}_j$ respectively, there must exist smooth functions $\alpha_{ij}(q)$, $q \in \mathcal{U}_i \cap \mathcal{U}_j$, also called [30,99] *transition functions*, such that:

$$| \phi(q) >_i = \exp[i\alpha_{ij}(q)] \cdot | \phi(q) >_j \; ; \quad q \in \mathcal{U}_i \cap \mathcal{U}_j \tag{5.22}$$

The learned reader will have already recognized that, by associating a phase (i.e. an element of $U(1)$) smoothly on each "patch" U_i and by defining a set of transition functions on overlapping patches what we are constructing is actually a fiber bundle, and indeed *a principal $U(1)$ bundle* [30,99] *over* M. The local bases $| \phi >_i$ will allow us to define *local sections* of the bundle. The latter will be trivial iff it admits of a *global* section, i.e. iff the phase of $| \phi(q) >$ can be defined globally and in a smooth way over all of M.

Consider now a curve:

$$\gamma: [0,1] \to M \ , \ \gamma = \gamma(s) \ , \ 0 \leq s \leq 1 \tag{5.23}$$

on the base manifold M. We shall adopt here the following definition of parallel transport along the curve γ:

Definition: Given a normalized vector $| \psi(0) >$ in the ray spanned by $| \phi(\gamma(0)) >$, and hence:

$$| \psi(0) > \, = \exp[i\eta(0)] \cdot | \phi(\gamma(0)) > \tag{5.24}$$

a *parallel transport prescription* along γ is given whenever we can associate with $| \psi(0) >$ a *unique* normalized vector $| \psi(s) >$ in the ray spanned by $| \phi(\gamma(s)) >$:

$$| \psi(s) > \, = \exp[i\eta(s)] \cdot | \phi(\gamma(s)) > \tag{5.25}$$

such that, besides preserving normalization:

$$< \psi(s) \,|\, \psi(s) > = 1 \qquad (5.26)$$

(which is implicit in (5.25)), the parallel transport procedure obeys:

$$\eta(0) \rightarrow \eta(0) + \lambda \Rightarrow \eta(s) \rightarrow \eta(s) + \lambda \qquad (5.27)$$

The property (5.27) is often referred to as the property of *equivariance* [30] of the parallel transport procedure w.r.t. the action of the structure group $U(1)$ on the fibers.

The normalization condition (5.26) implies: $\text{Re}\{ < \psi \,|\, d\psi/ds > \} = 0$. We claim that *the same condition on the imaginary part of* $< \psi \,|\, d\psi/ds >$, *i.e. altogether:*

$$< \psi \,|\, \frac{d\psi}{ds} > = 0 \qquad (5.28)$$

uniquely defines a parallel transport procedure. Indeed, Eq. (5.28) implies:

$$< \psi(s) \,|\, \psi(s+\delta s) > = 1 + O((\delta s)^2) \qquad (5.29)$$

Let: $|\psi(s+\delta s) > =: |\psi > + |\delta\psi >$, and assume there is a second procedure, satisfying (5.28) but leading to, say: $|\psi(s+\delta s) > = |\psi > +$

$|\delta\psi>$'. As normalization is preserved in both cases:

$$|\psi> + |\delta\psi>' = \exp[i\mu] \cdot (|\psi> + |\delta\psi>) \qquad (5.30)$$

for some $\mu \in \mathbb{R}$, with $\mu \rightarrow 0$ when $\delta s \rightarrow 0$. Hence, to lowest order:

$$|\delta\psi>' = i\mu|\psi> + |\delta\psi> \qquad (5.31)$$

But Eq. (5.28) implies : $<\psi|\delta\psi> = <\psi|\delta\psi>' = 0$. Hence, multiplying by $<\psi|$:

$$i\mu<\psi|\psi> = 0 \Leftrightarrow \mu = 0 \qquad \square \qquad (5.32)$$

Let's go back now to Eq. (5.25). We have just proved that, given $\eta(0)$, $\eta(s)$ is unique. In view of Eq. (5.27), there is no loss of generality in setting $\eta(0)=0$. Differentiating Eq. (5.25), we obtain:

$$\frac{d}{ds}|\psi(s)> = i\frac{d\eta}{ds}|\psi(s)> + \exp[i\eta(s)]\frac{d}{ds}|\phi(s)> \qquad (5.33)$$

Eq. (5.28) implies then that $\eta(s)$ is determined by the differential equation:

$$\frac{d\eta}{ds} = i<\phi(s)|\frac{d\phi}{ds}> \quad , \quad \eta(0)=0 \qquad (5.34)$$

This is the basic equation defining parallel transport on our bundle. Explicitly, in the domain of a chart:

$$\left|\frac{d\phi}{ds}\right> = \left|\frac{\partial\phi(q)}{\partial q^i}\right> \cdot \frac{d\gamma^i}{ds} \; ; \quad q=\gamma(s) \tag{5.35}$$

We can define then the one-form:

$$\omega=: \; <\phi|\; d\phi> \tag{5.36}$$

which reads, in coordinates:

$$\omega= \; <\phi|\; \frac{\partial\phi}{\partial q^i}>\; dq^i \tag{5.37}$$

Denoting then by $\dot\gamma$ the tangent vector at γ:

$$<\phi|\; \frac{d\phi}{ds}> = \omega(\dot\gamma) \tag{5.38}$$

Under a change of coordinates (i.e. a change of basis) in the bundle:

$$|\phi(q)> \; \rightarrow \; \exp[i\lambda(q)] \; |\phi(q)> \tag{5.39}$$

(5.36) transforms as:

$$\omega \rightarrow \omega + id\lambda \tag{5.40}$$

(note that, as $<\psi\,|\,\psi>\,=1$, ω itself is **purely imaginary**).

Remark: The choice of $|\,\phi>$ as been done, and can be done in general, only patchwise on open sets. It follows that ω is **not** a global one-form on M. It is rather a **collection of one-forms**, each defined on one of the sets $\mathfrak{U}_i$ of an open covering of M. Any two such forms (cfr. Eqns. (5.22), (5.39) and (5.40)) are connected by a "gauge transformation" of the form (5.40). The set of the patchwise defined ω's is what is called a set of **local connection forms** on the base manifold. It can be proved [30] that there is a **unique** (connection) one-form, globally defined **on the bundle** (and not on the basis) for any given family of local connection one-forms. It is actually easy to see that, calling z ($|z|=1$) an element of the fiber over q, the one-form:

$$\tilde{\omega}=:\ \omega-\mathrm{i}\frac{\mathrm{d}z}{z} \tag{5.41}$$

is in principle only patchwise defined, but is actually "gauge invariant", and hence is a global one-form. On any one of the $\mathfrak{U}_i$'s, the pull-back of $\tilde{\omega}$ via the corresponding local section yields back ω. In the previous Section we have been working directly in terms of the connection one-form on the bundle. However, the construction (5.41) is essentially what was employed in Sect. 2.2 to write down Eqns. (2.26-7). In particular, the first term on the r.h.s. of Eq. (2.26) is precisely a local connection one-form on the open set V_0 defined there.

Let now γ be a loop in M:

$$\gamma = \gamma(s) \ ; \ \ \gamma(0) = \gamma(1) = q \in \mathsf{M} \tag{5.42}$$

The total phase change of the parallel-transported wave function along γ will be then:

$$\eta(\gamma) = i \int_0^1 ds\ \omega(\dot\gamma) \equiv i \int_\gamma \omega \tag{5.43}$$

If γ bounds a surface Σ contained in M we have, by Stokes' theorem:

$$\eta(\gamma) = i \int_\Sigma \Omega \tag{5.44}$$

where:

$$\Omega = d\omega = \frac{1}{2} \left\{ < \frac{\partial\phi}{\partial q^i} | \frac{\partial\phi}{\partial q^j} > - < \frac{\partial\phi}{\partial q^j} | \frac{\partial\phi}{\partial q^i} > \right\} dq^i \wedge dq^j \tag{5.45}$$

is the **curvature two-form** associated with ω. Note that Ω is gauge-invariant, hence a globally defined two-form on M.

I will now turn to the Adiabatic Theorem of Quantum Mechanics [73]. I will not prove the theorem here, but simply state it.

Let's assume that the conditions stated in the Introduction hold, namely that the eigenvalue $E_0(q)$ is nondegenerate and separated from the rest of the spectrum by an everywhere finite gap along a path (not necessarily a loop) in M, and that the path is traced in the time interval $[0,T]$. Defining the scaled time variable $s=t/T$ and the evolution operator as $\hat{U}_T(s)$, the latter obeys the equation:

$$i\,\frac{d\hat{U}_T}{ds}= T\hat{H}(s)\cdot\hat{U}_T(s) \; ; \quad \hat{U}_T(0)= \mathbb{I} \tag{5.46}$$

Define also the projection operator projecting onto the (one-dimensional) eigenspace of the eigenvalue $E_0(s)$ by $\hat{P}(s)$. Then, the theorem states that:

$$\lim_{T\to\infty} \hat{U}_T(s)\cdot\mathbb{P}(0)= \mathbb{P}(s)\cdot[\lim_{T\to\infty} \hat{U}_T(s)] \tag{5.47}$$

The remarkable discovery of Berry's [18] was that, under an adiabatic evolution, the wave function changes as in Eq. (5.25), and that the phase η *is governed precisely by Eq.* **(5.34)**. Eq. (5.43) (or (5.44) whenever appropriate) gives then *Berry's phase* [18,21,22], a gauge invariant phase that develops under an adiabatic cyclic evolution: *cyclic adiabatic evolutions are equivalent to parallel transport of the wave function along loops in parameter space.*

The close similarity of the underlying mathematical structures leads us to conclude that the Pancharatnam phase analyzed in Sect. 5.2 is indeed a classical precursor of Berry's phase.

As stressed by Berry himself, manifestations of the quantum adiabatic phase had been found already in the late Seventies in the realm of Molecular Physics. More recently, Zak [114] has shown that quantum adiabatic phases arise also in the effective-Hamiltonian theory of shallow impurity states in semiconductors.

We will close this Chapter by analyzing a perhaps simpler and certainly more pedagogical example. Namely, let us show [18] how Berry's phase can be related to the Aharonov-Bohm effect [3,76], and hence also to the important problem of inequivalent quantizations discussed previously in Ch.3.

Consider a particle inside an infinite potential well which can move along a circle. Calling x the angular coordinate of the particle and Δ the angular width of the well, the motion will be limited by: $0 \leq$ x-R $\leq \Delta$, $\Delta < 2\pi$, $0 \leq$ R $< 2\pi$. R, the lower edge of the well, will play the rôle of the adiabatic parameter of the problem. Let the Hamiltonian be:

$$\hat{H} = \frac{1}{2m} [\frac{\hbar}{i} \frac{d}{dx} - \delta]^2 \tag{5.48}$$

δ can be interpreted as the vector potential of a flux line carrying a total flux $\Phi = 2\pi c \delta/e$, piercing the circle through the center. We are modelling thus the Aharonov-Bohm effect.

For every fixed R, the instantaneous eigenfunctions of (5.48) are:

$$\phi_n(x)=0, \quad x-R \notin [0,\Delta] \; ; \tag{5.49-a}$$

$$\phi_n(x)= \frac{2}{\Delta} \sin[\frac{\pi n}{\Delta}(x-R)]\exp[i\frac{\delta}{\hbar}(x-R)] \;, \quad x-R \in [0,\Delta] \tag{5.49-b}$$

with energies:

$$E_n= \frac{1}{2m}\left(\frac{\pi\hbar n}{\Delta}\right)^2 \tag{5.50}$$

We obtain therefore:

$$<\phi_n\mid d\phi_n> = -i\frac{\delta}{\hbar}\,dR \tag{5.51}$$

(note that, R being an angle, dR is a closed but not exact one-form). For a complete revolution around the circle, we then obtain Berry's phase as:

$$\eta= 2\pi\frac{\delta}{\hbar} \equiv 2\pi\frac{\Phi}{\Phi_0} \;; \quad \Phi_0= \frac{hc}{e} \tag{5.52}$$

i.e. *Berry's phase is precisely the Aharonov-Bohm phase that an electron accumulates by circling around a solenoid carrying a nonzero magnetic flux.*

CH.6. ELECTRONS IN A MAGNETIC FIELD, AND A CURSORY LOOK AT THE QUANTUM HALL EFFECT.

6.1. INTRODUCTION.

The discovery of the (integer and fractional) Quantum Hall Effect has led to a renewed interest in the study of the dynamics of charged particles in a static magnetic field.

That the addition of a magnetic field produces nontrivial characteristics of the quantum mechanical description of the motion of charged particles had been known actually for quite some time. In a sense, the crucial new feature is that the vector potential enters in an essential way into the definition of both the Lagrangian and the Hamiltonian. This is true of course at the classical level as well but, while the classical equations of motion involve only the Lorentz force and hence the magnetic field, the quantum mechanical amplitude turns out to depend on the choice of the vector potential itself, i.e. it becomes gauge-dependent in an essential way. It is therefore in the presence of a magnetic field that, in order to enforce gauge independence, we are led forcedly to interpret the wavefunctions as **sections** of a $U(1)$ bundle [12-15,110] over the configuration space. Such an interpretation of the wavefunction (see also Ch.3) becomes essential whenever a global definition of the vector potential is not possible, as it happens in the presence of magnetic poles [35,36].

At a more familiar and elementary level, the continuous energy spectrum of free-electron motion collapses, so-to-speak, into a discrete spectrum of harmonic-oscillator-like levels, the well known **Landau levels** (see Sect. 6.2 below).

Features that are even more interesting emerge when one considers the effect of a magnetic field not on free electrons, but on electrons moving in a lattice. In zero field, the discrete group of lattice translations (which is $\mathbb{Z} \times \mathbb{Z}$ for motion in a plane) can be implemented as a unitary, Abelian group commuting with the Hamiltonian, and this leads to the usual Bloch form for the wavefunctions. For nonzero magnetic field, however, the canonical way of realizing the translations as unitary operators on the Hilbert space of states allows only for a *ray* representation [75] of the set of translations. Loosely speaking, we obtain in this way what looks like a nonabelian group [66,75].

When the magnetic flux per unit cell of the lattice is a rational multiple of the elementary fluxon $\Phi_0 = hc/e$, it is possible [66] to extract an Abelian subgroup out of the whole set of translation operators, and this leads to generalized Bloch conditions on an enlarged lattice. The study of these so-called "magnetic translation groups" was initiated in the early Sixties by J. Zak (see [75] for a review).

The band structure that even a simple tight-binding model can develop when a uniform magnetic field is switched on is also unusual. D.R. Hofstadter [55] showed that the energy spectrum depends in a highly nonanalytic way on the flux per unit cell, and that it displays an amazing complexity, including self-similarity and a Cantor-set structure.

The magnetic field produces also singularities of the phase of the wavefunctions, and hence zeroes of the wavefunctions themselves, both in real and in reciprocal space (in the so-called "magnetic Brillouin zone", see below Sect. 6.3). The latter in turn lead to topological quantization of the Hall conductance [66,75,83,97].

Although strictly related to the results we have just mentioned, the Quantum Hall Effect (both integer and fractional) takes place however in a somewhat different scenario. Indeed, the integer effect (IQHE) shows up [75,83] in highly disordered samples, where the very concepts of lattice structure and Brillouin zones loose their meaning. Nonetheless, quantization of the Hall conductance emerges from this scenario as well as a topological effect. The fractional effect (FQHE) seems to be instead mainly a consequence of strong electron correlations [68,69,83]. Through the study of the FQHE we will make contact again with the subject of anyons discussed in Ch.3. Indeed, there is by now convincing theoretical evidence that elementary excitations above the ground state in the FQHE carry a *fractional* electric charge. If this is so one can conclude, via an Aharonov-Bohm-type argument [6,8,88], that they must obey fractional *statistics* as well, and this leads to the rather well-supported prediction of a *hierarchy* of FQHE states [83,113].

After some preliminaries in Sects. 6.2-4, the IQHE will be discussed in Sect. 6.5, the FQHE and fractionally-charged elementary excitations in Sects. 6.6-7.

6.2. PRELIMINARIES.

We shall consider here free electrons with charge -e-
moving in a **uniform** and **constant** magnetic field $\vec{B}$ directed along the
z-axis of a Cartesian reference frame: $\vec{B} = B\vec{k}$ ($|\vec{k}|=1$). The electron
dynamics along the z-axis is of course trivial and decouples from that in
the orthogonal plane(s), so we shall consider only motions in the (x,y)
plane. The Hamiltonian for free electrons in the presence of the field $\vec{B}$
can be written as:

$$\mathcal{H} = \frac{\vec{\Pi}^2}{2m} \tag{6.1}$$

where:

$$\vec{\Pi} =: -i\hbar\nabla + e\frac{\vec{A}}{c} \tag{6.2}$$

is the **kinetic** momentum ($\vec{\Pi} = m \cdot \vec{v}$, and -$\vec{v}$- is the quantum operator
corresponding to the velocity), and -$\vec{A}$- is the vector potential: $\vec{B} = \nabla \times \vec{A}$.

Working out the commutation relations, we find:

$$[\Pi_x, \Pi_y] = -i\hbar m\Omega \tag{6.3}$$

where: $\Omega =: eB/mc$ is the **cyclotron (or Larmor) frequency.** Defining
next:

$$a =: \frac{1}{\sqrt{2m\hbar\Omega}} \left(\Pi_x - i\Pi_y \right) \tag{6.4}$$

one easily finds the commutation relation:

$$[a, a^\dagger] = 1 \tag{6.5}$$

and:

$$\mathcal{H} = \hbar\Omega \left[a^\dagger a + \tfrac{1}{2} \right] \tag{6.6}$$

The spectrum of $\mathcal{H}$ is then that of a simple harmonic oscillator of proper frequency Ω, and the result is of course gauge independent.

The ground state manifold is spanned by the solutions of the equation:

$$a \, | \, 0 > \; = 0 \; ; \; < 0 \, | \, 0 > \; = 1 \tag{6.7}$$

and the (normalized) excited states are given by:

$$| \, n > \; = (n!)^{-\frac{1}{2}} \, (a^\dagger)^n \; | \, 0 > \tag{6.8}$$

Of course, there will be a "tower" of excited states (labeled by the integer -n-) for each independent solution of Eq. (6.7).

We discuss now the detailed structure of the solutions in some relevant gauges, namely:

i) The **Landau** gauge, defined by:

$$\vec{A}= B(0,x,0) \tag{6.9}$$

In this gauge, $\hat{p}_y \equiv$ -i$\hbar \nabla_y$ is a constant of the motion, and can be diagonalized along with $\mathcal{H}$. Adopting periodic boundary conditions in the y-direction (assumed to be of total length -L-), the eigenfunctions will be given by:

$$\Psi_{n,k}(x,y)= \frac{1}{\sqrt{L}}\exp[-iky]\phi_{n,k}(x) \;; \quad k=\frac{2\pi p}{L}; \;\; p \in \mathbb{Z} \tag{6.10}$$

and the effective (one-dimensional) Hamiltonian acting on the reduced wavefunction $\phi_{n,k}(x)$ is:

$$\mathcal{H}_{eff}= -\frac{\hbar^2}{2m}\frac{d^2}{dx^2} + \frac{1}{2}m\Omega^2(x-l^2k)^2 \tag{6.11}$$

where -l- is the **magnetic length** :

$$l= \sqrt{\frac{\hbar c}{eB}} \tag{6.12}$$

leading to:

$$\phi_{n,k}(x) = N_n \exp[-(x-l^2k)^2/2l^2] \cdot H_n(x/l-kl) \qquad (6.13)$$

with N_n a normalization factor, and H_n the n-th Hermite polynomial. For further use, let's remark that the magnetic length satisfies: $2\pi l^2 B = hc/e$, where the r.h.s. is a ratio of universal constants.

ii) The *symmetric* gauge, defined by:

$$\vec{A} =: \tfrac{1}{2} \vec{B} \times \vec{r} \qquad (6.14)$$

The Hamiltonian exhibits then cylindrical symmetry, and the z-component of the angular momentum:

$$\hat{L}_z = -i\hbar \left(x \cdot \frac{\partial}{\partial y} - y \cdot \frac{\partial}{\partial x} \right) \qquad (6.15)$$

is conserved. Introducing complex variables $z, \bar{z}$, $z =: x+iy$, the ground-state equation becomes:

$$[\frac{\partial}{\partial \bar{z}} + \frac{z}{2l^2}] \Psi_0(z,\bar{z}) = 0 \qquad (6.16)$$

Setting then:

$$\Psi_0 =: \exp[-|z|^2/4l^2] \cdot \Phi_0 \qquad (6.17)$$

Eq. (6.16) becomes:

$$\frac{\partial}{\partial \bar{z}}\Phi_0 = 0 \tag{6.18}$$

Therefore, as far as the ground-state manifold is concerned, Φ_0 *can be a function of -z- alone.*

The simultaneous ground-state eigenfunctions of $\mathcal{H}$ and $\hat{L}_z$ are:

$$\Psi_{m,0}(x,y) = [(2l)^{m+1}\sqrt{\pi m!}\,]^{-1}\, z^m\, \exp[-\pi\, |z|^2/4l^2] \tag{6.19}$$

Remark. In the gauge $\nabla \cdot \vec{A}=0$ (which includes both (6.9) and (6.14)) the vector potential can be obtained as the two-dimensional "curl" of a scalar (cfr. Ch.4) **magnetic potential** ϕ:

$$A_x = -\partial_y\phi; \quad A_y = \partial_x\phi \tag{6.20}$$

For the symmetric gauge (6.14):

$$\phi = \frac{B}{4}|z|^2 \tag{6.21}$$

and, in units $\hbar=c=1$, the exponential in (6.17) equals $\exp[-e\phi]$. In this form, (6.17) has been generalized to doubly-periodic fields by Dubrovin and Novikov [39].

Counting of states (In the Landau gauge, but the same results obtain in the symmetric gauge).

The wave vector -k- is quantized according to:

$$k = \frac{2\pi p}{L} \; ; \; p \in \mathbb{Z} \tag{6.22}$$

States of the form (6.13) are Gaussians centered around : $x_n =: l^2 k$. If the sample extends, in the -x- direction, from $-W/2$ to $+W/2$ for some (finite) W, then:

$$-\frac{W}{2} < x_n < +\frac{W}{2} \tag{6.23}$$

and the total number of states for fixed -n- (i.e. the degeneracy of a Landau level) is given by:

$$N_B = \frac{LW}{2\pi l^2} \tag{6.24}$$

while the **density of states** (number of states per unit area) for the n-th level is given by:

$$n_B = [2\pi l^2]^{-1} \equiv \frac{eB}{hc} \tag{6.25}$$

Remarks :

i) In terms of the elementary fluxon: $\Phi_0 = \frac{hc}{e}$

$$N_B \equiv \frac{\Phi}{\Phi_0} \; ; \quad \Phi =: BLW \tag{6.26}$$

i.e., N_B *is the number of elementary fluxons threading the sample, and shall be always assumed to be an integer.*

ii) If the sample contains -n- electrons per unit surface, the **filling factor** (or "*filling fraction*") -ν- is defined by:

$$\nu =: \frac{n}{n_B} \equiv \frac{nhc}{eB} \tag{6.27}$$

Integer value of -ν- correspond then to (one or more) completely filled Landau levels. Note that one can also write the total number of electrons N as: $N = \nu \Phi / \Phi_0$.

6.3. THE CLASSICAL HALL EFFECT.

Let us make now a brief interlude, by considering the **classical** theory of the Hall effect. Consider a slab of conducting material in crossed (static) electric and magnetic fields $\vec{E}$ and $\vec{B}$ (see Fig. 20). Let us assume also for simplicity the absence of damping mechanisms for the electron motion. Then, if $|\vec{E}| < |\vec{B}|$, there is a **net** current flow at right angles with both $\vec{E}$ and $\vec{B}$, whose direction depends on the sign of the charges involved. If $\vec{E}=0$, we know what happens from elementary textbooks: the electrons move in circular orbits whose radius is the "Larmor radius" (v/Ω) $(v=|\vec{v}|$ is a constant of the motion). The effects of a nonvanishing electric field can be "boosted away" by performing a Lorentz boost along the y-axis with boost velocity:

$$v_D = \frac{cE}{B} \qquad (6.28)$$

(as $E \ll B$ in typical experiments, actually $v_D \ll c$, and the boost is actually a Galilei boost). The resulting trajectories are known as "trochoids", and are superpositions of circular and uniform drift motions. The net current is therefore (with -n- the (2D) carrier density):

$$J_H = nv_D e \equiv \frac{nec}{B} \cdot E \qquad (6.29)$$

thus defining the *Hall conductivity* :

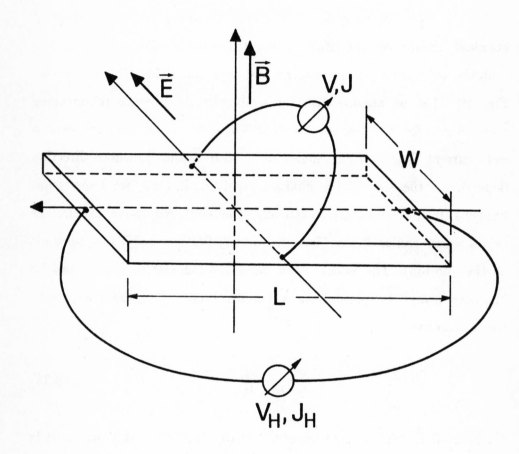

Fig. 20. A (theorist's) view of an experimental setup for the Quantum Hall Effect: $\vec{E}$, $\vec{B}$= applied electric and magnetic fields; W, L= sample sizes; V, J= "normal" parallel potential drop and current; V_H, J_H= Hall voltage and current.

$$\sigma_H =: \frac{nec}{B} \qquad (6.30)$$

In two space dimensions, the ratio: e^2/h has the dimensions of a conductivity, and: $\frac{nech}{Be^2} \equiv \nu$, the filling factor defined in Sect. 6.2. So, we obtain:

$$\frac{h}{e^2} \cdot \sigma_H = \nu \qquad (6.31)$$

Therefore, in "natural" units of e^2/h, $\sigma_H = \nu$, corresponding to the dotted line of Fig. 21.

6.4. BLOCH ELECTRONS IN A MAGNETIC FIELD.

Let us consider now electrons on a regular 2D lattice. For simplicity we shall consider here a rectangular Bravais lattice with primitive translation vectors $\vec{a}$ and $\vec{b}$ along the coordinate axes.

The electron Hamiltonian will be assumed to be invariant under the (discrete) group of lattice translations labeled by the vectors of the form:

$$\vec{R} = m\vec{a} + n\vec{b} ; \quad m,n \in \mathbb{Z} \qquad (6.32)$$

We will work here in the symmetric gauge (6.14). In the presence of a nonvanishing magnetic field, the generator [75] of translations in the

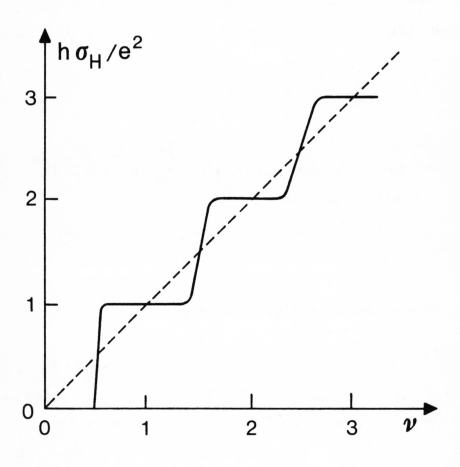

Fig. 21. A schematic view of the Integer Quantum Hall Effect (IQHE).

plane is not the canonical momentum $\vec{p}$ but rather:

$$\vec{\pi} = \vec{p} + \frac{e}{2}\,\vec{r} \times \vec{B} \tag{6.33}$$

Therefore, translations by lattice vectors of the form (6.32) are unitarily generated by the operators:

$$T_{\vec{R}} = \exp[i\vec{R}\cdot\vec{\pi}_{op}/\hbar] \; ; \; \vec{\pi}_{op} = \frac{\hbar}{i}\nabla + \frac{e}{2}\,\vec{r}\times\vec{B} \tag{6.34}$$

The $T_{\vec{R}}$'s will commute with the Hamiltonian, but not among themselves, and, in fact [66,75]:

$$T_{\vec{a}}\cdot T_{\vec{b}} = T_{\vec{b}}\cdot T_{\vec{a}}\,\exp[2\pi i\Phi/\Phi_0] \tag{6.35}$$

where: $\Phi = Bab$ is the magnetic flux through the unit cell and Φ_0 the flux quantum.

Let now Φ/Φ_0 be a rational number: $\Phi/\Phi_0 = p/q$ with p, q mutually prime integers. Then, we can enlarge the unit cell to a *magnetic* cell with sides qa and b, restricting at the same time the lattice translations to the subgroup indexed by vectors of the form:

$$\vec{R}' = mq\vec{a} + n\vec{b} \; ; \; m,n \in \mathbb{Z} \tag{6.36}$$

Now, the $T_{\vec{R}'}$'s will commute among themselves, and can be used to impose the usual Bloch conditions on the wavefunctions in the form:

$$T_{q\vec{a}}\Psi = \exp[ik_1 qa]\Psi$$

$$(6.37)$$

$$T_{\vec{b}}\Psi = \exp[ik_2 b]\Psi$$

where the crystal momentum $\vec{k} \equiv (k_1, k_2)$ ranges now in the *magnetic Brillouin zone*: $|k_1| \leq \pi/qa$, $|k_2| \leq \pi/b$.

The wavefunctions can be written in Bloch form as:

$$\Psi_{\vec{k},\alpha}(\vec{r}) = \exp[i\vec{k}\cdot\vec{r}]\, u_{\vec{k},\alpha}(\vec{r}) \qquad (6.38)$$

where: $\vec{r} \equiv (x,y)$ and α is a band index. Eqns. (6.37) imply then:

$$u_{\vec{k},\alpha}(x+qa,y) = \exp[-i\pi py/b]\, u_{\vec{k},\alpha}(x,y)$$

$$(6.39)$$

$$u_{\vec{k},\alpha}(x,y+b) = \exp[i\pi px/qa]\, u_{\vec{k},\alpha}(x,y)$$

By writing the u's as:

$$u_{\vec{k},\alpha}(\vec{r}) = |u_{\vec{k},\alpha}|\exp[i\theta_{\vec{k},\alpha}(\vec{r})] \qquad (6.40)$$

we can view, for fixed $\vec{k}$ and α, the wavefunction $u_{\vec{k},\alpha}(\vec{r})$ as a complex order parameter field over the unit cell similar to those describing, say, He^4 and superconductors that have been discussed in Ch. 1. Now, it is immediate to prove from Eqns. (6.39) that, by circling counterclockwise around the boundary $\mathcal{C}$ of the magnetic unit cell, the total phase change of (6.40), which must be an integer multiple of 2π, is given precisely by $2\pi p$, i.e. that [66,75]:

$$\frac{1}{2\pi} \int_{\mathcal{C}} d\theta_{\vec{k},\alpha} = p \tag{6.41}$$

We can apply now the results of Ch. 1, and conclude that *there must be a distribution of singularities of the phase inside the unit cell with total winding number* p. Moreover, according to the discussion of Sect. 2.5, we can conclude also that to each singularity of the phase there must correspond a zero of the wavefunction. If each zero is counted as many times as the winding number of the associated phase singularity, *for rational flux* $\Phi = (p/q)\Phi_0$ *there must be* (at least) p *zeroes of the wavefunction(s) inside the magnetic unit cell.*

The Hall conductivity for Bloch electrons in a magnetic field has been calculated by Thouless et al. [97] in a paper that has played a crucial rôle in the understanding of the topological nature of the IQHE. The main result of [97] (but see also [66]) can be rephrased as follows. Define, inside the magnetic Brillouin zone (MBZ) the one-form:

$$\mathbb{A}_\alpha = \int d^2r \, u_{\vec{k},\alpha}(\vec{r})^* d_k u_{\vec{k},\alpha}(\vec{r}) \equiv \, <u_{\vec{k},\alpha} \, | \, d_k \, | \, u_{\vec{k},\alpha}> \qquad (6.42)$$

where the integral is over the magnetic unit cell and d_k is the exterior differential operating on the $\vec{k}$-dependence of the wavefunctions. Using Linear Response Theory and the Kubo formula [75], Thouless et al. [97] showed that the Hall conductivity $\sigma_{H,\alpha}$ of a full band is given by:

$$\sigma_{H,\alpha} = \frac{e^2}{h} \int_{C'} \left(\mathbb{A}_\alpha / 2\pi i \right) \qquad (6.43)$$

where C' is the boundary of the MBZ.

The integral on the r.h.s. of (6.43) represents again the *total phase change of the wavefunction* (divided by 2π) but, this time, *around the boundary of the* MBZ. As such, it must be again an integer, and the Hall conductivity of a full band turns out to be an *integer* multiple of the fundamental unit e^2/h. Once again (see the previous discussion) *a nonvanishing Hall conductivity $\sigma_{H,\alpha}$ forces the wavefunctions to develop* (at least) $(h/e^2)\sigma_{H,\alpha}$ *zeroes inside the* MBZ.

By using Stokes' theorem [30], (6.43) can be rewritten also as:

$$\sigma_{H,\alpha} = \frac{e^2}{h} \int_{MBZ} (\mathbb{F}/2\pi i) \; ; \quad \mathbb{F} \equiv d_k \mathbb{A} \qquad (6.44)$$

All of the above results stress the bundle interpretation of Quantum Mechanics that has been discussed in Sect. 6.1 and, more extensively, in Chapts. 3 and 5. The one-form $\mathbb{A}_\alpha$ of Eq. (6.42) appears as a *connection form on a* $\mathbb{U}(1)$ *bundle over the* MBZ, while the two-form (6.44) is the associated *curvature* form.

As to the base manifold (i.e. the MBZ), it can be viewed as a manifold with boundary [75], and then both (6.43) and (6.44) apply. However, as the wavefunctions $u_{\vec{k},\alpha}$ are periodic in reciprocal space, the MBZ can be viewed also [66,75] as a two-torus $\mathbb{T}^2$, i.e. as a compact manifold without boundary. In such a case, the Hall conductivity must be defined directly by Eq. (6.44), and its integrality results from the integral in (6.44) being the *first Chern class* [30,99] associated with the curvature of the bundle.

As discussed in Sect. 6.1, however, the QHE takes place in a somewhat different scenario than that of Bloch electrons in a perfect lattice. In the following Sections, we shall discuss how the topological interpretation of the QHE survives the effects of disorder that are present in the real samples exhibiting it.

6.5. THE INTEGER QUANTUM HALL EFFECT.

The by now famous results of von Klitzing, Dorda and Pepper [101] are schematically represented by the full line of Fig. 21. Their main features are:

i) Well defined *plateaux* (at low temperatures) centered around *integer* fillings, and

ii) *Integer* values (in units of e^2/h) of the plateau Hall conductivity:

$$\sigma_H = n\,\frac{e^2}{h}\;;\;\; n \in \mathbb{Z} \tag{6.45}$$

The systems in which this *Integral Quantum Hall Effect (IQHE)* shows up (at high fields and low temperatures) are:

i) Essentially 2D systems (MOSFET's and heterostructures), and also, especially

ii) Disordered systems (this being due to the preparation procedures).

We inquire now briefly on the effect of impurities and disorder on the structure of Landau levels. In a perfectly pure sample, and for $\vec{E}=0$, the density of states is composed of a series of δ-functions centered at the position of the Landau levels $E_n = (n + 1/2)\hbar\Omega$. It can be shown easily [75,83] that the introduction of the electric field lifts completely the degeneracy of the Landau levels, modifying the spectrum from E_n to:

$$E_{n,k} = (n + \tfrac{1}{2})\, \hbar\Omega + eE l^2 k \qquad\qquad (6.46)$$

The original sharp Landau levels are then broadened into bands centered at $(n+1/2)\hbar\Omega$ and of total width eEW, corresponding to the potential drop across the sample. As, under typical experimental conditions (that is for high magnetic fields), $eEW \ll \hbar\Omega$, the bands are well separated from each other (the separation remaining roughly of the order of $\hbar\Omega$).

The effect of impurities and disorder turns out to be that of splitting *localized states* off each band (from both above and below), which tend to fill in the gap between adjacent subbands, until the (qualitative) picture of the density of states looks as in Fig. 22. As long as the Fermi level lies within a tail of localized states, there are no electron states available to contribute to the conduction besides those which already do, and hence one should expect σ_H=const. under such conditions. On the contrary, the conductivity is expected to rise sharply when the Fermi level sweeps a new set of extended states. *The above picture for the density of states explains the plateaux, but of course not the integer values of $\sigma_H \cdot (h/e^2)$.*

Thouless and coworkers (see [83] for a review) have given perhaps the most important contribution towards a complete understanding of the integral quantization of the conductivity in the

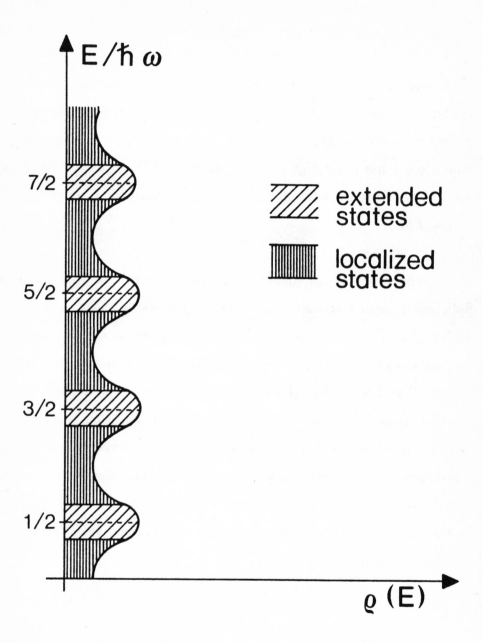

Fig. 22. The actual density of states $\rho(E)$ in a disordered sample, showing tails of localized states.

IQHE, while Avron, Seiler and Simon [11] performed the topological analysis in a completely convincing way. We omit details here, but only quote the main results of the above authors.

By evaluating the conductivity via Linear Response Theory and the Kubo formula, Thouless and coworkers proved a result that has proved to be crucial for all subsequent developments, namely that *the Hall conductivity $\sigma_H(\vec{r})$ at any point $\vec{r}$ inside the sample is a* <u>*local*</u> *quantity, depending only on the properties of the medium in an immediate neighborhood of $\vec{r}$, with corrections dying away exponentially on the scale of the magnetic length.* The main consequence of this result is that the Hall conductivity is **essentially insensitive to the boundary conditions** one imposes at the edges of the physical samples. The Hall conductivity one measures in actual experiments is given by the sample-averaged conductivity:

$$\sigma_H = \frac{1}{A} \int d^2r \, \sigma_H(\vec{r}) \quad ; \quad A = LW \tag{6.47}$$

Under the assumption that **the Fermi level lies in a tail of** *localized states* , Thouless and coworkers found, for the sample-averaged Hall conductivity:

$$\sigma_H = \sum_n f(\epsilon_n) \, \sigma_{H,n} \tag{6.48}$$

where the ϵ_n's are the single-particle energy eigenvalues, -n- is a Landau-level index, f(.) is the Fermi function, and:

$$\sigma_{H,n} = \frac{i\hbar e^2}{A} \sum_{m \neq n} \frac{(v_x)_{nm}(v_y)_{mn} - (v_y)_{nm}(v_x)_{mn}}{(\epsilon_n - \epsilon_m)^2} \tag{6.49}$$

Recall that the velocity operator $\vec{v}$ is given by:

$$\vec{v} = \frac{1}{m} \left[\frac{\hbar}{i} \nabla - \frac{e\vec{A}}{c} \right] \tag{6.50}$$

Adding a pure gauge field, i.e. letting

$$\nabla \to \nabla + \frac{i}{\hbar} \vec{\lambda} \; ; \quad \vec{\lambda} = \text{const.} \tag{6.51}$$

the gauge field can be reabsorbed by redefining the boundary conditions as:

$$\Psi(x+W) = \exp[-i \frac{\lambda_x W}{\hbar}] \Psi(x)$$

$$\Psi(y+L) = \exp[-i \frac{\lambda_y L}{\hbar}] \Psi(y) \tag{6.52}$$

The wavefunctions depend then *periodically* (with period 2π) on the angles:

$$\alpha =: \frac{\lambda_x W}{\hbar} \quad , \quad \beta =: \frac{\lambda_y L}{\hbar} \tag{6.53}$$

Moreover:

$$v_x = \frac{\partial \mathcal{H}}{\partial \lambda_x} = \frac{W}{\hbar} \frac{\partial \mathcal{H}}{\partial \alpha} \; ; \quad v_y = \frac{\partial \mathcal{H}}{\partial \lambda_y} = \frac{L}{\hbar} \frac{\partial \mathcal{H}}{\partial \beta} \tag{6.54}$$

We obtain therefore:

$$\sigma_{H,n} = 2\pi i \frac{e^2}{h} \sum_{m \neq n} \frac{1}{(\epsilon_m - \epsilon_n)^2} \left\{ <n | \frac{\partial \mathcal{H}}{\partial \alpha} | m> <m | \frac{\partial \mathcal{H}}{\partial \beta} | n> - (\alpha \Leftrightarrow \beta) \right\}$$

$$\tag{6.55}$$

Avron, Seiler and Simon [11] showed further that (6.55) can be recast into the more compact form:

$$\sigma_{H,n} = 2\pi i \frac{e^2}{h} \left\{ <\frac{\partial n}{\partial \alpha} | \frac{\partial n}{\partial \beta}> - <\frac{\partial n}{\partial \beta} | \frac{\partial n}{\partial \alpha}> \right\} \tag{6.56}$$

In view of the stated independence of the Hall conductivity on the boundary conditions, one can replace (6.56) by its average on the boundary conditions, thereby obtaining eventually:

$$\sigma_{H,n} = \frac{e^2}{h} \int_0^{2\pi} \int_0^{2\pi} \frac{\Omega_n}{2\pi i} \tag{6.58}$$

where the two-form Ω_n is given by:

$$\Omega_n = d\mathbb{A}_n \; ; \quad \mathbb{A}_n = \; < dn \mid n > \tag{6.59}$$

In this form, Ω_n is easily recognized to be *the curvature two-form of a connection $\mathbb{A}_n$ on a principal $U(1)$ bundle over the base space parametrized by the angles α and β. $(h/e^2) \cdot \sigma_{H,n}$ is then the first Chern number [30] associated with the above connection, and is hence an integer.* As, at low temperatures: $f(\epsilon) \simeq \theta(\epsilon_F - \epsilon)$, ϵ_F being the Fermi energy, the total Hall conductivity will be an integer as well.

From the form of the connection, one can further conclude that $(h/e^2) \cdot \sigma_{H,n}$ *can also be viewed as the Berry phase associated with an adiabatic circuit along the boundary ($\alpha=0$ or 2π and/or $\beta=0$ or 2π) of the set of boundary conditions labeled by the angles α and β.*

6.6. THE FRACTIONAL QUANTUM HALL EFFECT.

We turn now to a brief account of the *Fractional Quantum Hall Effect* (**FQHE**). This effect was discovered about two years after the IQHE. It occurs in highly pure samples, at lower temperatures ($\sim 1^0$K) and higher fields (of about 15 Teslas) than the IQHE, again with plateaux and exact quantization of the Hall conductivity, namely:

$$\sigma_H = \nu \frac{e^2}{h} \tag{6.60}$$

but with $\nu = p/q$, p and q being relatively prime integers, Further, apart from some recent claims, q is an *odd* integer, with typical values $\nu = 1/3$ (the best known experimental value), 2/5, 2/7 etc. We shall concentrate here on the (by now) "classic" $\nu = 1/3$ case.

The status of the topological analysis of the effect is definitely not as clear as it is for the IQHE, although some authors [113] have argued that, *if* there is a gap of extended states at $\nu = p/q$, then the ground state should be q-fold degenerate, and that an extension of the topological analysis should lead to fractional quantization.

It seems however that the most widely accepted "paradigm" for the structure of the ground state, due to Laughlin [68,69], leaves very little elbow room for degeneracy, and therefore that the explanation for the experimentally found fractional quantization calls for some kind of entirely different theoretical approach.

We shall discuss now briefly Laughlin's approach to the description of the ground state and of the (low lying) excited states appropriate for the FQHE. Laughlin's ground state wavefunction is a variational function, meant to describe an assembly of **strongly correlated** electrons in the lowest (n=0) Landau level. The description of the electron system departs then considerably from that appropriate to the IQHE which, we recall, is in terms of essentially independent (quasi) particles subject to impurity scattering. Here it is believed that Coulomb correlations play a dominant rôle, and that **gaps at rational fillings are entirely due to many body effects.**

With reference to what has already been discussed, the single particle wavefunctions for the lowest Landau level, and in the symmetric gauge, can be written as:

$$\psi_{p,0}(z) = \tilde{N}_{p,0} \exp[\,|z|^2/4l^2\,] \, (\partial_{\bar{z}})^p \exp[-\,|z|^2/2l^2\,] \qquad (6.61)$$

with $\tilde{N}_{p,0}$ a normalization factor. If the lowest Landau level is entirely filled ($\nu=1$), the ground state wavefunction is given, in the absence of correlations, by the Slater determinant of the single particle wavefunctions (6.61) for $p=0,...N_B-1$. The latter turns out to be essentially a Vandermonde determinant, and we have, calling Ψ_0 the ground-state wavefunction, and $-N^*-$ a normalization factor:

$$\Psi_0 = N^* \prod_{i<j} (z_i - z_j) \exp\left[-\sum_{k=1}^{N} |z_k|^2/4l^2\right] \qquad (6.62)$$

Note that the (z-component of) the angular momentum operator can be written as:

$$\hat{L} = \hbar \sum_{i=1}^{N} \left(z_i\, \partial_{z_i} - \bar{z}_i \partial_{\bar{z}_i} \right) \qquad (6.63)$$

so that:

$$\hat{L}\, \Psi_0 = \frac{N(N-1)}{2}\hbar\Psi_0 \qquad (6.64)$$

that is, Ψ_0 is an eigenfunction of the total angular momentum with eigenvalue $N(N-1)\hbar/2$.

In the first [68] of many seminal papers, Laughlin proposed a trial wavefunction of the form:

$$\Psi = N^* \prod_{i<j} f(z_i - z_j) \exp\left[-\sum_k |z_k|^2/4l^2\right] \qquad (6.65)$$

where the "a priori" unknown function f should satisfy the following requirements:

i) Having to describe electrons confined to the lowest Landau level, -f- should be *a function of the z's alone* (and not of the $\bar{z}$'s);

ii) As a consequence of Pauli's principle, f should be *odd under any interchange i↔j* , and

iii) -f- should be also *an eigenfunction of the total angular momentum* (which commutes with the total Hamiltonian in the case of interactions such as the Coulomb interaction as well).

All this boils down to -f- being of the form:

$$f(\zeta) = \zeta^m , \quad \text{-m- an } \textbf{\textit{odd}} \text{ integer} \tag{6.66}$$

and we end up with the final form of Laughlin's wavefunction, namely:

$$\Psi = N^* \prod_{i<j} [\frac{z_i - z_j}{l}]^m \exp[-\sum_k |z_k|^2/4l^2] \tag{6.67}$$

The wavefunction Ψ of Eq. (6.67) is now an eigenfunction of the total angular momentum with eigenvalue $N(N-1)m\hbar/2$.

The probability density can be written as:

$$|\Psi|^2 = \text{const.} \cdot \exp[-\beta\Phi] \tag{6.68}$$

where: $\beta = 1/m$, and:

$$\Phi =: -2m^2 \sum_{i<j} ln\left|\frac{z_i-z_j}{l}\right| + \frac{1}{2}m \sum_i \frac{|z_i|^2}{l^2} \qquad (6.69)$$

In this form, $|\Psi|^2$ can be viewed as the *classical probability distribution of a one-component plasma* (OCP) at "temperature" $\beta=1/m$, the plasma being made up of particles of charge Q= m which:

i) Repel each other via logarithmic interactions (the natural form of the Coulomb interaction in two space dimensions) and

ii) interact with a fixed background of charge of opposite sign and charge density:

$$\sigma = (2\pi l^2)^{-1} \equiv n_B \qquad (6.70)$$

The value of m is then fixed by the requirement of *electrical neutrality* of the system. If there are n particles per unit surface area, then we must have: $n \cdot m = n_B$, and, as $\nu = (n/n_B)$, we eventually find:

$$\nu = \frac{1}{m} \qquad (6.71)$$

establishing the link between the filling fraction and the exponent m.

6.7. FRACTIONALLY-CHARGED QUASIPARTICLES AND THE HIERARCHY OF QUANTUM HALL STATES.

Elementary excitations above Laughlin's ground state, in the nature of quasiparticles or quasiholes, can be created in the following manner. Consider again, for the moment, the free-electron system, with the single-particle eigenfunctions (6.61). It can be proved [75] that, if one pierces the plane at $z=0$ with an "infinitely thin" solenoid and varies adiabatically the flux inside the solenoid from zero to $\Phi_0 = hc/e$ (that is by one flux quantum), the result is an adiabatic map of state p into state (p+1) ((p-1) if the flux is varied by $-\Phi_0$, with the state labeled by $p=0$ being mapped into the next Landau level). If one removes now the solenoid (remember that, if $\Phi/\Phi_0 =$ integer, this can be done with the aid of a nonsingular gauge transformation), the net result of this "Gedankenexperiment" will be to leave a hole at the place where the solenoid was (a particle if $\Phi = -\Phi_0$). By switching on the interaction adiabatically, the state will evolve into an excited state of the *full* Hamiltonian containing just one quasihole (this is in the same spirit as the description of quasiholes (or quasiparticles) in Landau's Fermi Liquid theory).

Generalizing this argument to a quasihole centered at an arbitrary point z_0, Laughlin [68,69] proposed an (approximate) wavefunction for one hole at z_0 in the form:

$$\overset{+}{\Psi}_{z_0} = N_+ \, A_{z_0} \, \Psi \; ; \; A_{z_0} =: \prod_{i=1}^{N} (z_i - z_0) \qquad (6.72)$$

The adjoint operator:

$$A_{z_0}^{\dagger} =: \prod_{i=1}^{N} \left(\frac{\partial}{\partial z_i} - \frac{\bar{z}_0}{l^2} \right) \qquad (6.73)$$

will in turn create a quasiparticle at z_0. Note that A_{z_0} and $A_{z'_0}^{\dagger}$ do **not** commute for $z_0 \neq z'_0$. This is connected with the fact that the quasiparticles have finite size, and hence the processes of creating or destroying them at different points are not independent processes.

It has been proved [83] that:

i) Comparison with numerical calculations gives excellent agreement between (6.72) and the exact wavefunctions (calculated for systems with a finite number of (actually few) particles, of course).

ii) One can actually construct (see the contribution of F.D.M. Haldane in [83]) model Hamiltonians for which both Laughlin's ground state and excited states are *exact* eigenstates.

We want to discuss now both the charge and statistics of Laughlin's quasiparticles.

One way of measuring the charge of a particle is to carry it adiabatically around a solenoid enclosing a total flux -Φ-. Then, if we denote by -e^*- the particle's charge, the **adiabatic (Berry) phase** its wavefunction will acquire is given by:

$$\Delta\gamma = \frac{e^*}{\hbar c} \int \vec{A} \cdot \vec{dl} \equiv 2\pi \frac{e^*}{e} \cdot \frac{\Phi}{\Phi_0} \qquad (6.74)$$

where $\vec{A}$ is the vector potential, and the integral is taken along the adiabatic circuit. The phase (6.74) can be measured in an Aharonov-Bohm-type interference experiment. On the other hand, if $|\psi(t)>$ is the instantaneous wavefunction, Berry's phase can be calculated (see Ch.5) from the parallel-transport equation:

$$\frac{d\gamma}{dt} = i <\psi(t)|\frac{d\psi}{dt}> \qquad (6.75)$$

Consider now the process by which a quasihole located initially at z_0 is adiabatically dragged along a circle of radius $R=|z_0| \gg 1$ centered at the origin. The adiabatic circuit will enclose a flux Φ determined, at filling ν, by the condition:

$$<n>_R = \nu \frac{\Phi}{\Phi_0} \qquad (6.76)$$

where $<n>_R$ is the average number of the electrons inside the disk of radius -R-. Employing the wavefunction (6.72), we obtain at once:

$$\frac{d}{dt}\Psi^+_{z_0} = \sum_i \frac{d}{dt} \, ln(z_i - z_0(t)) \, \Psi^+_{z_0} \qquad (6.77)$$

and hence:

$$\frac{d\gamma}{dt} = i \, <\Psi^+_{z_0}| \frac{d}{dt}\sum_i \ln (z_i - z_0(t))| \Psi^+_{z_0}> \qquad (6.78)$$

Introducing the average electron density in the state (6.72) as:

$$\rho_+(z) = \, <\Psi^+_{z_0}| \sum_i \delta(z - z_i) | \Psi^+_{z_0}> \qquad (6.79)$$

we obtain:

$$\frac{d\gamma}{dt} = i \int dxdy \, \rho_+(z) \frac{d}{dt} \ln (z - z_0(t)) \qquad (6.80)$$

We expect $\rho_+(z)$ to be of the form:

$$\rho_+(z) = \rho_0 + \delta\rho_+(z) \qquad (6.81)$$

with: $\pi R^2 \rho_0 = \, <n>_R$, and $\delta\rho_+$ a localized correction with spatial extent of the order of the size of the quasihole. The total variation of $\ln(z-z_0)$ will be $2\pi i$ if $|z| < R$, zero otherwise, and we obtain:

$$\Delta\gamma = -2\pi < n >_R + \text{finite corrections} \qquad (6.82)$$

and, as $< n >_R \propto R^2$:

$$\Delta\gamma = -2\pi < n >_R = -2\pi\nu \frac{\Phi}{\Phi_0} \qquad (6.83)$$

in the limit of large R's. Comparing with Eq. (6.62), we see that **the quasiholes carry a fractional charge** given by:

$$e^* = -\nu e \qquad (6.84)$$

The same will be true for the quasiparticles, with a charge of opposite sign and same magnitude.

Consider now two quasiholes located at some positions z_a and z_b respectively, with $|z_a - z_b| = R$, and the process of dragging hole "b" adiabatically around hole "a". The fact that hole "a" carries a charge $-\nu e$ can be interpreted as implying that **exactly ν electrons** have been removed from the (big) disk of radius R centered at z_a. The same considerations as before apply, but for the fact that $< n >_R$ must be substituted by $< n >_R - \nu$, and **an extra phase (of statistical origin)**

$$\Delta\gamma' = 2\pi\nu \qquad (6.85)$$

is accumulated in the process. It should be clear from the previous discussion that the extra phase is accumulated steadily during the adiabatic motion. If we consider now the process of *exchange* of two quasiholes, a process which can be accomplished by letting each of them make a π turn around the other, *the total phase change will be $\pi\nu$.* For $\nu=1$ (mod. 2π) the quasiholes (as well as the quasiparticles) will be fermions, while they will be bosons for $\nu=0$ (again mod. 2π). For *fractional* fillings, however, they will obey *fractional (or "anyon") statistics.*

Elementary excitations above Laughlin's ground state are to be considered as identical particles moving in a 2D manifold (essentially $\mathbb{R}^2$). Their (inequivalent) scalar quantizations are associated with the one-dimensional unitary representations of the corresponding braid group ($\mathbb{B}_n$ for -n- particles), which we know from Ch. 3 to be classified by an angle θ, $0 < \theta < 2\pi$. Not only does the example of the FQHE show a concrete physical realization of "braid statistics", it shows also that, among all the "kinematically" equally possible values of θ, the actual dynamics (and the stability of the ground state) acts to select a specific value of θ, namely: $\theta=\pi\nu$, where $1/\nu$ is an odd integer.

One of the most striking consequences of the existence of fractionally-charged quasiparticles in the FQHE is that they predict a whole *hierarchy* [83]of FQHE states, many of whose members have been observed in actual experiments. Let us summarize then the arguments leading to the prediction of such a hierarchy. Let us recall, first of all, that numerical calculations prove [83] that the ground state at $\nu=1/m$ is separated by a **finite gap** from the low-lying excited states.

Laughlin's state is therefore an **incompressible** state [69], and fractional Hall conductivity results from fractional filling.

Moving away from $\nu=1/m$ will result in the production of quasiparticles or quasiholes. Just as in the case of the IQHE, one should expect that, as long as the concentration of elementary excitations is small enough, the latter will be "pinned" by impurities, and this will explain qualitatively the plateaus in the Hall conductivity. As the concentration increases, elementary excitations will behave however as a gas of (fractionally!) charged particles, and they will eventually "condense" in a Laughlin-type "liquid" described by a wavefunction of the form, qualitatively, of (6.67), exhibiting again incompressibility, quantized Hall conductance and so on.

Let us make now these considerations a bit more quantitative. Let the wavefunction for an assembly of N quasiparticles be of the form:

$$\Psi'(z_1,...,z_N)=\prod_{j<k=1}^{N}(z_j-z_k)^{m_1}\exp\left(-\frac{e}{e^*l^2}\sum_k|z_k|^2\right) \quad (6.86)$$

for some integer m_1. By comparison with (6.12) we see that e^*l^2/e is the magnetic length appropriate to particles with charge e^*. The integer m_1 must be determined by the requirement that, when two elementary excitations are circled around each other, the wavefunction should acquire a phase of: $\exp[i\pi m_1]=\exp[i\pi\alpha/m]$, where $\alpha=1$ for quasiholes, $\alpha=-1$ for quasiparticles. This leads [113] to:

$$m_1 = 2p_1 + \frac{\alpha}{m} \qquad (6.87)$$

with p_1 an integer.

One can develop again an OCP analogy for the wavefunction (6.86). Charge neutrality condition will fix the density of elementary excitations in this case as well. To be precise, quasiholes will move in a background of positive charge of density:

$$\sigma^* = \frac{e^*}{e}\, \sigma \qquad (6.88)$$

where σ is given again by (6.70). Electrical neutrality will require then the density n_1 of quasiholes to obey:

$$\frac{n_1 m_1}{n_B} = \frac{\sigma^*}{n_B} \Rightarrow n_1 = \frac{1}{mm_1} \qquad (6.89)$$

As each elementary excitation carries a charge α_1/m, the **electron** filling factor in the state (6.86) will be:

$$\nu_1 = \nu - \frac{\alpha_1 n_1}{m} \equiv \frac{2p_1}{2mp_1 + \alpha_1} \qquad (6.90)$$

Thus, for $m=\nu^{-1}=3$ and $p_1=1$,say, we obtain $\nu_1=2/7$ for quasiholes, and $\nu_1=2/5$ for quasiparticles respectively, which are both observed fractions.

The procedure outlined here can be iterated and leads, as already mentioned, to the prediction of a whole hierarchy of FQHE states, whose filling factors are determined recursively by a continuous fraction expression [83,113]. As mentioned at the beginning of this Section, the prediction of the hierarchy of FQHE states is the most striking theoretical evidence of the *physical reality of anyons in the* **FQHE**.

REFERENCES

[1] ABRAHAM, R.; MARSDEN, J.E.; RATIU, T.:
"Manifolds,Tensor Analysis and Applications."
2nd Ed., Springer Verlag, 1988.

[2] AFFLECK, I.: Nucl.Phys. B257, 397 (1985).

[3] AHARONOV, Y.; BOHM, D.: Phys. Rev. 115, 485 (1959)

[4] AITCHISON, I.J.R.: Acta Phys. Polonica B18, 207,
(1987)

[5] AITCHISON, I.J.R.: Phys. Scripta T23, 12 (1988)

[6] AROVAS, D.P.: "Topics in Fractional Statistics". In
Ref. [88]

[7] AROVAS, D.P.; SCHRIEFFER, J.R.; WILCZEK, F.:
Phys. Rev. Letters 53, 722 (984)

[8] AROVAS, D.P.; SCHRIEFFER, J.R.; WILCZEK, F.;
ZEE, A.: Nucl. Phys. B251, 117 (1985)

[9] ARTIN, E.: Ann. Math. 48, 101, 643 (1947)

[10] AVRON, J.E.; SEILER,R.: J. Geom. and Phys. 1, 13,
(1984)

[11] AVRON, J.E.; SEILER, R.; SIMON, B.: Phys. Rev.
Letters 51, 51 (1983)

[12] BALACHANDRAN, A.P.: "Classical Topology and Quantum Phases: Quantum Mechanics" In Ref. [32].

[13] BALACHANDRAN, A.P.: "Classical Topology and Quantum Phases". In Ref. [27].

[14] BALACHANDRAN, A.P; MARMO, G.; SKAGERSTAM, B.S.; STERN,A. : " Gauge Symmetries and Fibre Bundles". Springer-Verlag, 1983.

[15] BALACHANDRAN, A.P; MARMO, G.; SKAGERSTAM, B.S STERN, A.: "Classical Topology and Quantum States". World Scientific, 1991.

[16] BALACHANDRAN, A.P.; ERCOLESSI, E.; MORANDI,.G.; SRIVASTAVA, A.M.: Int. J. Mod. Phys. $\underline{B4}$, 2057, (1990).

[17] BALACHANDRAN, A.P.; ERCOLESSI. E.; MORANDI, G.; SRIVASTAVA, A.M.: "Hubbard Model and Anyon Superconductivity". World Scientific, 1990.

[18] BERRY, M.V.: Proc. Roy. Soc. $\underline{A392}$, 45 (1984)

[19] BERRY, M.V.: J. Phys. $\underline{A18}$, 15 (1985)

[20] BERRY, M.V.: J. Mod. Optics $\underline{34}$, 1401 (1987).

[21] BERRY, M.V.:"The Quantum Phase, Five Years Later". In Ref.[88].

[22] BERRY, M.V.: "Quantum Adiabatic Anholonomy". In Ref. [27].

[23] BIRMAN, J.S.: "Braids, Links and Mapping Class Groups". P.U.P., 1975.

[24] BLOORE, F.J.: "Configuration Space of Identical Particles". In Ref. [47].

[25] BOTT, R.; CHERN, S.S: Acta Math. $\underline{114}$, 71 (1965)

[26] BOTT, R.; TU, L.: "Differential Forms in Algebraic Topology". Springer-Verlag, 1982.

[27] BREGOLA, M.; MARMO, G.; MORANDI, G. (Eds.): "Anomalies, Phases, Defects...", Bibliopolis, Naples,1990.

[28] BURKE, W.L.: "Applied Differential Geometry". Cambridge Univ.Press., 1985.

[29] CHERN, S.S; SIMONS, J.: Acta Math. $\underline{99}$, 48 (1974).

[30] CHOQUET-BRUHAT, Y.; DEWITT MORETTE, C.: "Analysis, Manifolds and Physics". Part I, 2nd Ed.: North-Holland, 1982. Part II: North-Holland, 1989.

[31] CRAMPIN, M.; PIRANI, F.A.E.: "Applicable Differential Geometry". C.U.P., 1986.

[32] DEFILIPPO,S.; MARINARO, M.; MARMO, G.; VILASI,G. (Eds.): "Geometrical and Algebraic Aspects of Nonlinear Field Theory". North-Holland, 1989.

[33] DESER, R.; JACKIW, R.; TEMPLETON, S.: Ann. Phys. $\underline{140}$, 37 (1982).

[34] DEWITT, B.S.; STORA, R. (Eds.): "Relativity, Groups and Topology II". Les Houches 1983. North-Holland, 1984.

[35] DIRAC, P.A.M.: Proc. Roy. Soc. A133, 60 (1931)

[36] DIRAC, P.A.M.: Phys. Rev. 74, 817 (1948).

[37] DOWKER, J.S.: "Selected Topics in Topology and Quantum Field Theory". Austin Lectures 1979, unpublished.

[39] DUBROVIN, B.A.; NOVIKOV, S.P: Sov. Phys. JETP 52, 511, (1980).

[40] FELSAGER, B.: "Geometry, Particles and Fields". Odense Univ. Press, 1981.

[41] FEYNMAN, R.P.; HIBBS, A.R.: "Quantum Mechanics and Path Integrals". McGraw-Hill, 1965.

[42] FINKELSTEIN, D.; RUBINSTEIN, J.: J.Math. Phys. 9, 1762, (1968).

[43] FLANDERS, H.: "Differential Forms". Ac. Press, 1963.

[44] FRADKIN, E.; STONE, M.: Phys. Rev. B38, 7215 (1988).

[45] FRANCAVIGLIA, M.: "Elements of Differential and Riemannian Geometry". Bibliopolis, Naples, 1988.

[46] FRÖLICH, J.; GABBIANI, F.: Revs. Math. Phys. 2, 251, (1990).

[47] GARCIA, P.L.; PEREZ-RENDON, A.; SOURIAU, J.M. (Eds.): "Differential Geometric Method in Mathematical Physics". Springer-Verlag, 1979.

[48] GODDARD, P.; MANSFIELD, P.: Rep. Prog. Phys. 49, 725 (1986).

[49] GOVINDARAJAN, T.R.; SHANKAR, R.: Mod. Phys. Letters $\underline{A4}$, 1457 (1989).

[50] GUTIERREZ, M.: "Lecture Notes on Braids". University of Naples, 1991, unpublished.

[51] HALDANE, F.D.M.: Phys. Rev. Letters $\underline{50}$, 1153 (1983)

[52] HALDANE, F.D.M.: Phys. Rev. Letters $\underline{51}$, 605 (1983)

[53] HANNAY, J.H.: J.Phys. $\underline{A18}$, 221 (1985)

[54] HILTON, P.J.: "An Introduction to Homotopy Theory". Cambridge .Univ.Press, 1961.

[55] HOFSTADTER, D.R.: Phys. Rev. $\underline{B14}$, 2239 (1976)

[56] HORVATHY, P.A.; MORANDI,G.; SUDARSHAN, E.C.G: Nuovo Cim. $\underline{D11}$, 2239 (1989).

[57] HU,S.T.: "Homotopy Theory". Ac. Press, 1959.

[58] IMBO, T.D.; SUDARSHAN, E.C.G.: Phys. Rev. Letters 60, 48 (1988).

[59] ISHAM, C.J.: "Topological and Global Aspects of Quantum Theory". In Ref. [34].

[60] ISHAM C.J.: "Modern Differential Geometry for Physicists". World Scientific, 1989.

[61] JACKIW, R.: "Topological Investigations of Quantized Gauge Theories". In Ref. [34].

[62] JACKIW, R.: Int. J. Mod. Phys. $\underline{A3}$, 285 (1988).

[63] JACKIW,R.: Ann. Phys. $\underline{201}$, 83 (1990).

[64] JACKIW, R.; PI, S.Y.: Phys. Rev. $\underline{D42}$, 3500 (1990).

[65] KLEINERT, H.: "Path Integrals in Quantum Mechanics, Statistic and Polymer Physics" . World Scientific, 1990.

[66] KOHMOTO, M.: Ann. Phys. 160, 343 (1985).

[67] LAIDLAW, M.G.G.; MORETTE-DEWITT,CC.: Phys. Rev. D3, 1375 (1971).

[68] LAUGHLIN, R.B.: Phys. Rev. Letters 50, 1395 (1983).

[69] LAUGHLIN, R.B.: "Elementary Theory. The Incompressible Quantum Fluid". In Ref. [83].

[70] LAUGHLIN, R.B.: Science 242, 525 (1988).

[71] LEINAAS, J.M.; MYRHEIM, J.: Nuovo Cim. B37, 1, (1977).

[72] MERMIN, N.D.: Revs. Mod. Phys. 51, 531 (1979). Reprinted in Ref. [27].

[73] MESSIAH, A.: "Mecanique Quantique". Dunod, Paris, 1959.

[74] MILNOR, J.: Comm. Math. Helv. 32, 215 (1957).

[75] MORANDI, G.: "Quantum Hall Effect". Bibliopolis, Naples, 1988.

[76] MORANDI, G.; MENOSSI, E.: Eur. J. Phys. 5, 49, (1984).

[77] NAKAHARA, M.: "Geometry, Topology and Physics". A. Hilger, 1990.

[78] NASH, C.; SEN, S.: "Topology and Geometry for Physicists". Ac. Press, 1983.

[79] OLARIU, S.; POPESCU, I.I.: Revs. Mod. Phys. 57, 339, (1985).

[80] PAK, N.K.; PERCACCI, R.: Nucl. Phys. B188, 355 (1981).

[81] PANCHARATNAM, S.: "Generalized Theory of
Interference and its Applications". In: Collected
Works of S. Pancharatnam. Oxford Univ.
Press,1975.

[82] PESHKIN, M.; TONOMURA, A.: "The Aharonov-Bohm
Effect". Springer-Verlag, 1989.

[83] PRANGE, R.E.; GIRVIN, S.M. (Eds): "The Quantum
Hall Effect". 2nd Edition. Springer-Verlag, 1990.

[84] RAJARAMAN,R.: "Solitons and Instantons". North-Holland,1982.

[85] RICHTMYER, R.D.: "Principles of Advanced
Mathematical Physics". Springer-Verlag, vol.I, 1978;
vol. II, 1981.

[86] SAMUEL, J.; BHANDARI, R.: Phys. Rev. Letters 60,
2339 (1988)

[87] SCHULMAN, L.S.: "Techniques and Applications of
Path Integration". J. Wiley & Sons, 1981.

[88] SHAPERE, A.; WILCZEK, F. (Eds.): "Geometric Phases
in Physics". World Scientific, 1989.

[89] SIMON, B.: Phys. Rev. Letters 51, 2167 (1983).

[90] SINGER, I.M.; THORPE, J.A.: "Lecture Notes on
Elementary Topology and Geometry". Scott,
Foresman & Co., 1967.

[91] SKYRME, T.H.R.: Proc. Roy. Soc. A260, 237 (1961).

[92] SKYRME, T.H.R.: Nucl. Phys. 31, 556 (1962).

[93] SORKIN, R.D.: "Classical Topology and Quantum
Phases: Quantum Geons". In Ref. [32].

[94] STEENROD, N.: "The Topology of Fibre Bundles".
Princeton Univ.Press, 1951.

[95] STONE, M: Int. J. Mod. Phys. B4, 1465 (1990).

[96] SUDARSHAN, E.C.G.: "Topology and Quantum Internal
Symmetries in Nonlinear Field Theories". In Ref. [32].

[97] THOULESS, D.J.; KOHMOTO, M.; NIGHTINGALE,
M.D.; den NIJS, M. : Phys. Rev. Letters 49, 405,
(1982).

[98] TOULOUSE, G. : J. Phys. Lett. (Paris) 38, L67 (1977).

[99] TRAUTMAN, A.: "Differential Geometry for
Physicists". Bibliopolis, Naples, 1984.

[100] VOLOVIK, G.E.: Phys. Scripta 38, 321 (1988).

[101] VON KLITZING, K.: Revs. Mod. Phys. 58, 519 (1984).

[102] VON WESTWENHOLZ, C.: "Differential forms in
Mathematical Physics". North-Holland, 1981.

[103] WESS, J.; ZUMINO, B.: Phys. Letters 37B, 95 (1971).

[104] WILCZEK, F.: Phys. Rev. Letters 48, 1144 (1982).

[105] WILCZEK, F.: Phys. Rev. Letters 49, 957 (1982).

[106] WILCZEK, F. (Ed.): "Fractional Statistics and Anyon
Superconductivity". World Scientific, 1990.

[107] WILCZEK, F. (Ed.): "Fractional Statistics in Action".
World Scientific, 1991.

[108] WILCZEK, F.; ZEE, A.: Phys. Rev. Letters 51, 2250,
(1983)

[109] WILCZEK, F.; ZEE, A.: Phys. Rev. Letters $\underline{52}$, 2111, (1984).

[110] WU, T.T.; YANG, C.N.: Phys. Rev. $\underline{D12}$, 3845 (1975).

[111] WU, Y,S.: Phys. Rev. Letters $\underline{52}$, 2103 (1984).

[112] WU, Y.S.: Phys. Rev. Letters $\underline{53}$, 111 (1984).

[113] WU, Y.S. : "Topological Aspects of the Quantum Hall Effect". Lectures given at the NATO Advanced Summer Institute on: "Physics, Geometry and Topology", Banff, Alberta, Canada, 1989. To appear.

[114] ZAK, J.: Phys. Rev. $\underline{B40}$, 3156 (1989).

[115] ZAKRZEWSKI, W.J.: "Low Dimensional Sigma Models". A. Hilger, 1989.

[116] ZUMINO, B.: "Chiral Anomalies and Differential Geometry". In Ref. [34].

Lecture Notes in Physics

For information about Vols. 1–365
please contact your bookseller or Springer-Verlag

New Series m: Monographs

R. Fernández, J. Fröhlich, A. D. Sokal

Random Walks, Critical Phenomena and Triviality in Quantum Field Theory

1992. XVII, 444 pp. 26 figs. (Texts and Monographs in Physics)
Hardcover ISBN 3-540-54358-9

H. Spohn

Large Scale Dynamics of Interacting Particles

1991. XI, 342 pp. 19 figs. (Texts and Monographs in Physics)
Hardcover ISBN 3-540-53491-1

E. H. Lieb

The Stability of Matter: From Atoms to Stars

Selecta of Elliott H. Lieb

Edited by W. Thirring,
University of Vienna

With a Preface by F. Dyson

1991. VIII, 565 pp.
Hardcover ISBN 3-540-53039-8

Springer-Verlag
Berlin
Heidelberg
New York
London
Paris
Tokyo
Hong Kong
Barcelona
Budapest